One Hundred Centuries
of Solitude

One Hundred Centuries of Solitude

Redirecting America's High-Level Nuclear Waste Policy

James Flynn, James Chalmers, Doug Easterling,
Roger Kasperson, Howard Kunreuther,
C.K. Mertz, Alvin Mushkatel, K. David Pijawka,
and Paul Slovic

with Lydia Dotto, Science Writer

Routledge
Taylor & Francis Group

LONDON AND NEW YORK

First published 1995 by Westview Press

Published 2019 by Routledge
52 Vanderbilt Avenue, New York, NY 10017
2 Park Square, Milton Park, Abingdon, Oxon OX14 4RN

Routledge is an imprint of the Taylor & Francis Group, an informa business

A CIP catalog record for this book is available from the Library of Congress.

ISBN 13: 978-0-367-28190-8 (hbk)
ISBN 13: 978-0-367-29736-7 (pbk)

*To Gilbert White with
respect, affection, and gratitude*

Contents

A New Management Approach, 61

Preface

Time is both the ally of high-level nuclear waste (HLNW) managers and the enemy. It is the ally because the radioactivity in elements and isotopes decreases with age, making the waste progressively less dangerous to human health and safety and the environment. This rate of radioactive decline varies, in some cases diminishing by half (the half life) in seconds, minutes, hours, days, weeks, months, or years. In other cases the decay process takes centuries or hundreds of thousands of years before the wastes are safe for human contact. The problem as now conceptualized for HLNW managers is simple to state if not easy to achieve. The HLNW needs to be secured in some fashion until it decays, by virtue of its physical nature, to safe levels. Another possible future solution, not currently available, might be to change the structure of HLNW through high-technology processing and thus decompose the waste into units with different and less lengthy radioactivity. Learning whether this processing is a future option will require patience and generous amounts of time for research.

Time is also the great enemy because it introduces tremendous uncertainties and the possibility of almost infinite variations and combinations for future events. While the past is certain—although it may be mysterious because we do not know all that has happened—the future is a storehouse of surprises and revelations. The further we attempt to peer into the future, the less distinct our sense of possibilities becomes. Uncertainty crowds out thoughts, like the terrible deity Cronus of Greek mythology, ready to devour the birth of rationality.

The political process for dealing with the problem of HLNW ultimately resulted in congressional enactment of the Nuclear Waste Policy Act (NWPA) in 1982. Congress decided that HLNW should be stored in a geologic repository deep underground. The regulatory process mandated by NWPA also established a performance standard for the permanent repository. It should securely store spent fuel rods from nuclear power reactors and certain wastes from the federal nuclear weapons complex for 10,000 years. In other words, the repository should impose on the wastes 100 centuries of solitude.

The trick (or perhaps we should say the task) is to find some way to store these wastes and to be assured that they can be sequestered for 10,000 years. The responsibility for developing the repository was assigned by Congress to the Department of Energy (DOE). DOE's ideas of what they were to do, who they were to convince, and what evidence would be needed for what issues and concerns is one major subject of this book. After more than a decade and numerous large and small restructurings of the program, the goal of providing a repository for HLNW is as far away as ever. Despite the Congress' 1987 selection of Yucca Mountain, Nevada, as the nation's only site to be studied as a repository location, much doubt remains. The Yucca Mountain project teeters along, constantly on the verge of collapse. There is a very significant possibility that the site will not meet the technical, scientific, and public acceptance standards needed for success.

Based on research conducted since 1986, this book describes basic issues and concerns that beset the Yucca Mountain project, explaining the nature of the problems and reviewing how other countries have approached HLNW issues. We recommend, in chapter 7, a new approach to public policy for HLNW management. This approach would require a major change in how HLNW facilities are evaluated; it would raise the essential social, cultural, psychological, and political issues to parity with technical issues; and it would address the basic issues of developing a decision process that achieves public acceptance and support for HLNW outcomes. In short, this new approach would address a set of legitimate issues in HLNW management that have not been adequately dealt with in past programs. This limitation of past efforts accounts for the repeated failures to solve the HLNW problem.

The title of this book is adapted from the novel *One Hundred Years of Solitude* by Gabriel García Márquez. His work of fiction presents a somewhat strange world in which the events of nature and the actions of people are overwhelmed with surprises and sometimes amazing outcomes. The village of Macondo, for example, disintegrates under a steady rain that starts suddenly and lasts for 4 years and 11 months. Then one June morning it begins to clear, and it does not rain again for 10 years. In the 100 years of Márquez's story, very little happens in nature or in human experience as it should, that is, as people intended and expected. How much more uncertain we are in facing the prospect of 100 centuries, even if our gaze is focused only on a mountainous block of rock in the Nevada desert.

James Flynn
Eugene, Oregon

Acknowledgments

The Yucca Mountain study team members who participated in writing this book have worked together for more than 8 years. The study team has produced well over 300 reports, presentations, articles, books, and chapters in books. I would like to record here my deep appreciation for the privilege of working with these researchers, including those study team members who did not participate in the drafting of this policy-oriented book, but whose more general contributions, I hope, are adequately represented. In particular I would like to recognize George Blankenship and Jim Williams of Planning Information Corporation, Denver, Colorado; Dr. Catherine Fowler, Department of Anthropology, University of Nevada, Reno; and Drs. Richard Krannich and Ronald Little, Department of Sociology, Utah State University.

The study team was fortunate to obtain the services of science writer Lydia Dotto of Peterborough, Ontario, Canada. We appreciate her completing a difficult task with great skill and with an admirable good nature throughout.

We would also like to thank Mr. Joseph Strolin, our project director, for his long-term support. Mr. Strolin's participation with the study team work over the years has been substantive as well as administrative, and he should be recognized for his important contribution to the many social science achievements the study team has made. We also would like to record our thanks to Mr. Robert Loux, Nevada Nuclear Waste Project Office (NWPO) executive director, Harry Swainston, Nevada deputy attorney general, and Steve Frishman and Bob Halstead who are associated with NWPO.

While personnel at NWPO, other state and local officials, and the majority of Nevada citizens oppose the Yucca Mountain project, it is our impression that they would support a federal program based on fairness, equity, and an acceptable public decision process. However, we do not pretend to speak for them in this book except as they presented their views in our social science research, and we do not wish to present our recommendations as coming from Nevada and its residents. This book is not an attempt to present a Nevadan argument about Yucca Mountain.

Over the years we benefited from formal and informal discussions with members of the Technical Review Committee established to oversee and advise the socioeconomic research. We would like to thank Gilbert White, Michael Bronzini, William Colglazier, Bruce Dohrenwend, Kai Erikson, Reed Hanson, Allen Kneese, Richard Moore, Edith Page, and Roy Rappaport. During the entire period of the study team work, we have also benefited from the contributions of Drs. William Freudenburg and John Gervers, who served as independent consultants to NWPO. Also, I would like to recognize Mary Elizabeth Flynn for her support and assistance.

Ms. Kari Nelson at Decision Research served as the in-house editor and was responsible for manuscript preparation, copy editing, preparation of the camera-ready text, and essential editorial services in all phases of the process of making this book a reality. Her contribution was essential and is greatly appreciated by the authors. Other support assistance was provided by Toni Daniels, Geri Hanson, and Kay Phillips. Working at Decision Research is an especially rewarding experience because of the generous support and cooperation that is always available from Don MacGregor, Sarah Lichtenstein, Bob Clemen, Steve Johnson, and Terre Satterfield. A special thanks goes to my friend and colleague, Robin Gregory, who always knows when I'm serious and when I'm not—often better than I know myself.

The preparation of this book was supported in part by a contract between Decision Research and NWPO with federal funds pursuant to the provisions of Public Law 97–425. The government has certain rights in this material. Any opinions, findings, conclusions, or recommendations expressed in this book are those of the authors and do not necessarily reflect the views of the federal government or NWPO.

J.F.

1

Overview of a Troubled Program

The United States is facing one of the greatest scientific and technological challenges in its history: devising a system for safely storing high-level nuclear waste (HLNW) that will endure for 100 centuries in the face of all uncertainties, mishaps, and surprises the future will undoubtedly bring.

This is also a great political and social challenge. Few people welcome the prospect of having these lethal wastes transported past their homes and businesses or stored near their communities. Consequently, the success of any HLNW management program is critically dependent not only on technical know-how but also on cooperation and good-faith dealings among multiple levels of government and their agencies, conspicuous fairness in distributing the costs and benefits of the program, and, most importantly, earning the trust and acceptance of a public predisposed to fear nuclear waste and suspect the motives and methods of the institutions managing them.

HLNW—primarily spent fuel rods from nuclear power plants—are currently stored in water pools next to reactors, but because these wastes remain dangerously radioactive for thousands of years, they must eventually be shipped to and sequestered within a more permanent facility. In the United States, as in several other countries, the plan is to bury the wastes deep underground in a repository shielded from the surrounding environment by a massive barrier of stable, impermeable rock.

The federal government has been searching for a suitable geologic site for more than two decades and, at present, its attention is focused on a single candidate site: Yucca Mountain, Nevada. However, disposal cannot be achieved until the Department of Energy (DOE), which manages the program, characterizes the site—that is, determines whether the geologic structure is capable of safely containing radiation for 10,000 years. The Environmental Protection Agency (EPA) has estimated it will take that long for the radioactivity of the wastes to decay to levels no greater than that emitted by natural uranium. The data DOE collects during site

1

characterization will be used in its application to the Nuclear Regulatory Commission (NRC) for a construction license.

Although this process is still in the early stages, the Yucca Mountain project has become mired in a political, financial, legal, and public relations quagmire from which it appears unlikely to be extricated. The schedule has already slipped several times; originally scheduled to open in 1998, the facility is unlikely to be built even by the current official target of 2010—a date that the federal General Accounting Office (GAO) estimates is too optimistic by 5 to 13 years. In addition, cost estimates have soared; more than $2 billion has already been spent on site characterization, and DOE has projected that the cost of these studies could eventually exceed $6 billion—a far cry from the original 1982 estimate of $60 million.

These problems—combined with a dismal track record in handling defense wastes—have been enough in themselves to create serious doubts about DOE's ability to get the job done at Yucca Mountain, but there is more. In fact, the repository program is under siege from all fronts. There are scientific uncertainties about the safety and suitability of the only site being considered, support for the program from Congress and the nuclear utilities is wavering, and the state of Nevada has mounted a fierce campaign of opposition, bolstered by evidence of grave public mistrust of the project and its managers within Nevada and throughout the United States.

Nevadans are rightly concerned that siting a repository just 90 miles north of Las Vegas could adversely affect a state economy dominated by the tourist industry and uniquely dependent on gaming and entertainment. Socioeconomic studies have shown that the mere association of such a facility with a community generates extremely negative images in the minds of local residents and visitors, creating a stigma effect that could have major social and economic consequences by reducing a community's attractiveness as a place to visit or live.

In fact, a strong case can be made that the proposed Yucca Mountain repository represents a huge gamble with Nevada's future economic well-being. Even in a state that thrives on betting, it's a gamble Nevadans would rather not take.

The combination of technical difficulties, scientific uncertainties, escalating costs, repeated management failures, state opposition, and public distrust has so compromised the federal government's HLNW management program that it forces one to consider whether the current approach should be abandoned altogether. Would it be wise at this stage to retrench and reevaluate the entire effort, before billions more are spent on a program that appears to have little hope of solving its problems or gaining the state cooperation and public confidence it needs to succeed?

How did things go so badly off the tracks? This situation was certainly not the outcome envisioned when Congress passed the Nuclear Waste Policy Act (NWPA) in 1982.

Nuclear Waste Policy Act and Its Amendments

From the beginning the United States HLNW management program has had a fractious and troubled history. NWPA attempted to address some of these long-standing problems by establishing a set of democratic principles and ethical rules to guide the process of selecting a repository site. Several of these principles and rules were designed to produce an equitable outcome by

- mandating the selection of two repositories, one in the west, where DOE had already done some site studies, and the other in the east, where most of the country's nuclear wastes are generated.
- ensuring that those who benefit from the repository pay for it. NWPA required fees charged on nuclear-generated energy to go into a Nuclear Waste Fund to finance repository development.
- providing monetary compensation to those who live near the repository.

Other provisions of NWPA were intended to ensure an equitable site selection process. They required DOE to

- provide information about all of the activities associated with selecting and building the repository, including its scientific data and analyses, to affected stakeholders, including state governments, Indian tribes, and the public.
- employ an objective selection process based on technical criteria. In addition, the decision process used to select or eliminate candidate sites was to be open to outside scrutiny, making it difficult for DOE to select a site on arbitrary or capricious grounds.
- consult and cooperate with affected states and Indian tribes before making key decisions (however, this provision did not give stakeholders authority to control the siting process).
- provide funds to state and/or affected Indian tribes to oversee DOE studies and conduct socioeconomic assessments.

NWPA also permitted any state designated to host a repository to file a notice of disapproval—essentially a veto of the site. Although this notice exceeded previous state authority over siting of a federal facility, it was a weak power because any veto could be overturned by Congress.

Finally, NWPA contained provisions to protect public health. It instructed EPA to set radiation exposure standards. EPA subsequently determined that a repository could cause no more than 1000 excess cancer deaths over its 10,000-year lifetime. DOE was to demonstrate how it would meet these standards, which would be used by the NRC to decide whether to grant a construction license.

NWPA appeared to many (though not all) observers as a reasonable political compromise—a good-faith effort to put together a successful siting process. The equity and public safety provisions were intended to ensure fairness in site selection and to make the eventual choice acceptable to those who would be directly affected. And, indeed, NWPA seemed to have succeeded—at least to the extent of attracting, in 1982, the support of most of the congressional representatives from states that were then identified as potential repository host sites.

Unfortunately, more than a decade after its passage, many of NWPA's most admirable provisions have been abandoned or violated by subsequent actions of DOE and Congress. Among many notable transgressions, DOE's 1986 decision to abandon the search for a second repository site in the east, claiming that the single western facility would meet the country's needs, stands out. In the midst of an election period, this decision was seen as a capitulation to the political interests of Congressional representatives from populous eastern states that were being considered for sites. In 1987 this decision was incorporated into the NWPA Amendments Act (NWPAA), which went a step further by halting all site characterization studies except those at Yucca Mountain—also widely viewed as a politically-driven move.

Since 1988 Congress and DOE have repeatedly cut federal funding for the state of Nevada's own studies, which have challenged the technical suitability of the Yucca Mountain site and documented public antipathy toward the project. Finally, the 1992 Energy Policy Act weakened the health-protection provisions of NWPA by requiring DOE to meet potentially less stringent standards for radiation exposure than EPA had planned to implement.

All of these actions suggest that DOE and Congress are intent on building the repository at Yucca Mountain; any attempt to identify the technically-best site was abandoned by 1986, and almost all of NWPA's provisions designed to ensure a fair siting process have been rescinded or severely compromised.

It is hardly surprising, therefore, that Nevadans have been vociferous in their objections to nearly every aspect of the Yucca Mountain project. Even before the 1987 NWPAA was passed, the state was locked in legal combat with DOE. A 1985 lawsuit, for example, forced DOE to give Nevada funds for oversight studies, as required under NWPA. Political opposition in the state intensified after passage of NWPAA, widely touted as the "Screw Nevada Bill," and in 1989 the legislature passed two resolutions opposing the Yucca Mountain project and a bill (AB222) that made it illegal to dispose of HLNW in the state.

Although not all of Nevada's legal maneuvers succeeded, they clearly demonstrate that the state has the ability and resolve to continue attacking and delaying the Yucca Mountain project, including filing an official notice of disapproval at the licensing stage. Yucca Mountain will, in short, continue to be the focus of a prolonged, litigious, and messy conflict.

The fractious battles with Nevada, the schedule delays, and the escalating cost estimates, along with continuing evidence of DOE's management failures, have all undermined support for the Yucca Mountain project within Congress and the nuclear industry. So uncertain is the timetable for DOE to begin accepting HLNW that, in 1992, then-Energy Secretary James Watkins recommended that Congress return to the utilities part of the money they had paid into the Nuclear Waste Fund so they could build interim above ground storage facilities (i.e., dry casks for storing wastes at reactor sites).

Complexity and Scientific Uncertainty: Doubts About Yucca Mountain

With so much attention focused on the Yucca Mountain project, it is important not to lose sight of the fact that the repository is only one component of a national HLNW management program that has already evolved into a vast, complex, multibillion-dollar enterprise involving an extraordinary diversity of players: federal and state governments and agencies, local governments, numerous review and advisory committees, national laboratories, nuclear utilities, waste shippers and carriers, technology manufacturers, and a host of private contractors. Currently, more than 2000 people work in the program, and simply keeping the operation up and running consumes more than half the program's funds.

Many difficult questions face the managers of this far-flung enterprise, and not just concerning the repository itself. What kind of transportation system will be used to move the wastes? What role should be played by interim storage facilities at reactor sites or at central, above ground monitored facilities from which wastes can be retrieved? What type of containers should be used? Will new disposal technologies become available? Will future generations be more—or less—accepting of HLNW disposal facilities?

Tackling these issues (and many more) and integrating the diverse components of the HLNW management program into a cooperative, smoothly-functioning, and efficient operation capable of winning the confidence of a wary public is a considerable challenge. DOE's lackluster performance to date does not inspire confidence, nor does the frequent turnover in the program's leadership (there have been no fewer than seven directors in the past decade, five of whom never got beyond the status of acting director). DOE was so slow in getting the Yucca Mountain site characterization program underway that it would not have been able to field its scientific team on schedule even if Nevada had cooperated with the siting effort.

Further complicating the situation is the degree of scientific uncertainty associated with the HLNW management program generally and the Yucca Mountain project in particular. Demonstrating that nuclear wastes can be successfully isolated for 10,000 years is a task that far exceeds our current ability to project the course of natural and, especially, human events with even a pretense of precision. This uncertainty is especially true in the case of a facility that is the first of its kind, with all the unknowns that innovation inevitably entails. Unanticipated interactions among wastes, storage casks, and surrounding environments and groundwater may also exist. Indeed, predictions about the future behavior of the geologic structure, the stored wastes, their containers, and the repository itself rest on many assumptions that are themselves uncertain to varying degrees and, according to a 1990 statement from the National Academy of Sciences, "will remain uncertain no matter how much additional information is gathered" (NRC/NAS, 1990, p. 2).

Moreover, serious questions are being raised about events that might occur at Yucca Mountain over the next 10,000 years that could result in the release of unacceptable levels of radiation. Both catastrophic geological episodes and human intrusion are worrying possibilities. One theory, for example, suggests that Yucca Mountain may be subject to upwelling of groundwater—water forced up by a major seismic event like an earthquake. Flooding of the repository could conceivably lead to the release of radiation-contaminated water. A DOE-funded review of this theory by the National Academy of Sciences concluded that such a catastrophic event was unlikely. However, subsequent research sponsored by the state of Nevada seems to support the possibility. Scientists also believe that gaseous radioactive carbon-14 will almost certainly be released from the repository because a significant portion of the carbon-14 in the spent fuel rods is expected to outlast the waste containers.

Questions about whether seismic events might breach the containment system also exist—a concern underscored by the 1992 occurrence of a 5.6 magnitude quake near the repository site and evidence that two volcanoes near Yucca Mountain erupted within the past 5000 years.

Finally, there is no way of predicting the extent to which future human intrusion—accidental or deliberate—will pose a threat to the integrity of the repository. Accidental intrusion could occur if knowledge of the location of the facility is somehow lost—not a far-fetched idea considering the social, climatic, and geological changes that will undoubtedly occur over 10,000 years. In fact, future urban development could conceivably pose a threat. In the centuries to come, people might make far greater use of the underground environment for transportation and settlement. It's even possible that the wastes might be intentionally mined, if future technological developments make this economically worthwhile.

It is often assumed that continued application of scientific research and risk assessment techniques will reduce the technical uncertainties associated with the HLNW disposal program. Unfortunately, while such assessments may provide helpful information about what events might occur and when, and what their ramifications might be, they have only a very limited ability to predict the probability and magnitude of the dangers that may threaten the HLNW repository over the next 10,000 years. The inherent level of uncertainty in the undertaking will continue to defeat the ability of such techniques to provide proof that a repository will be safe over such a long time. The experience with assessments for a repository show that such studies are just as likely to increase uncertainty as to reduce it. Therefore, complexity and uncertainty seem destined to be a continuing part of the agenda and, as a result, the repository program should be viewed more as an experiment than as an established technological undertaking.

Ordinary citizens and their elected leaders will ultimately weigh the best evidence science can provide and make judgments, based on social values, about whether to proceed in the face of uncertainty and risk. But this approach can itself lead to conflict, especially when technical experts and managers view the HLNW program as a strictly technical problem and not a social issue whose outcome should at least partly be determined by people's values. Overreliance on scientific and technical expertise has the effect of further removing from the decision-making process the public and stakeholders who ultimately bear the risks. This distancing, in turn, increases the uncertainty that reliance on experts will ultimately be accepted. Evidence of this unacceptance is found in the increasing militancy of those potentially affected by technological programs. These people demand to have their concerns heard and considered by authorities, and they simply reject attempts to override their concerns by recourse to expert opinion. These attitudes create a significant source of uncertainty for the repository program at Yucca Mountain: the now widely recognized loss of public trust in the institutions responsible for managing HLNW.

Public Attitudes Toward Nuclear Waste Issues:
Gambling with the Quality of Life in Nevada

The public most often reacts to the idea of having a HLNW facility located near their communities or in their state with fear, distrust, and fierce opposition. Only a few communities—usually those historically associated with other nuclear facilities such as power plants or weapons manufacturing—have shown any willingness to host a nearby repository. Elsewhere, people find HLNW materials to be the least acceptable of hazardous wastes.

These fears are exacerbated by a deep and growing distrust of government institutions and the nuclear industry and by doubts about their ability to manage large nuclear programs safely and efficiently. The loss of public confidence is particularly salient to the problem of HLNW disposal, not just because radioactive materials are involved, but because they must be isolated over such a long period of time. That this disposal is a challenge unprecedented in human experience only intensifies public misgivings instilled by past failures.

Sanguine reassurances from technical experts usually do little to counteract such skepticism; in fact, these assurances often reinforce the determination of stakeholders to gain greater access to, and control over, the decision-making and policy processes. This determination, in turn, can lead to conflict with managers of the nuclear enterprise, who often regard dealing with public concerns primarily as an irritating complication in need of careful, even manipulative, handling. Indeed, scientists, engineers, and technically-oriented managers are often baffled by what they perceive as scientifically unsophisticated emotionalism on the part of public. These managers generally find the fact that the public is rarely sympathetic to the technological fix approach to nuclear issues exceedingly difficult to come to grips with and often mistakenly believe that if they redouble efforts to educate the public with facts, public opposition will simply melt away.

This management approach completely disregards the reality that most people who are asked (or forced) to host a nuclear facility are apt to mix personal, value-laden criteria into their assessment of risks. These criteria include fears for their health and safety and concerns for the environment and property values, risks to social and cultural life, quality of life considerations, psychological factors (such as their tolerance for risk-taking), and value judgments (as to the relative weight given to risks and benefits). The public's value judgments may differ markedly from value judgments made by those in the nuclear industry.

The public also factors in how much they believe the information they are given and, especially, how much they trust the messengers providing the information. The potentially catastrophic consequences of nuclear accidents and the many uncertainties associated with technical risk assessment creates a public demand for a high level of demonstrated competence and trustworthiness in those mandated to operate a HLNW disposal program.

In this respect, the federal government—and DOE in particular—have had a rough time. DOE's difficulties stem in part from the fact that it was given two distinct mandates that often conflicted with each other: to develop and promote nuclear technology and to ensure the public's safety from radioactive hazards. Some critics believe DOE's actions have shown a distinct bias in favor of its programs to promote nuclear technology. Moreover, evidence of past deceptions and mismanagement has done little to enhance DOE's reputation with the public.

NWPA was intended to address some of these problems. Unfortunately, it assigned DOE responsibility for implementing the innovative provisions relating to state and community participation and input into the repository program. Apparently unprepared for this unfamiliar role, DOE did not seem to view proposed host states and communities as possible allies, and certainly not as partners, but, at best, as potential service providers for solving the nuclear industry's waste problems—and as probable adversaries.

This attitude, of course, did little to allay the concerns of those opposed to the Yucca Mountain repository. As we have seen, their objections were based partly on the substantial scientific uncertainties associated with the project, but there was more to their objections than that. In Nevada there was grave concern about potential stigma effects on the state's economy—a concern that proved to be more than justified, as a large number of socioeconomic studies sponsored by the Nevada Nuclear Waste Project Office (NWPO) make clear. These studies were designed to assess the public's perception of the HLNW disposal program, to determine how this perception affected their attitudes and social behaviors, and to evaluate potential consequences for the social and economic well- being of communities near HLNW facilities and transportation routes. This work provides important insights into how large technological ventures like a HLNW repository can, by their mere existence, stigmatize a community, even if nothing goes wrong with the facility itself.

These studies had two main goals: (1) to document attitudes of Nevada residents to the repository and (2) to explore how opinions about a repository at Yucca Mountain might affect the willingness of people outside Nevada to vacation in the state, move there for employment or retirement, or invest there. Between 1986 and 1994 more than 20 surveys were conducted in Nevada, southern California (the major source of visitors to Nevada), and nationwide. Following are the major findings.

- Most respondents found a HLNW repository by far the most undesirable kind of facility to live near, compared with chemical waste landfills, oil refineries, pesticide plants, and even nuclear power plants.
- Most people believed that rail and highway accidents will occur in transporting HLNW to the repository. They also expected problems resulting from earthquakes or volcanic activity, contamination of underground water, and accidents in handling wastes during burial operations.
- A large majority of people said a state that does not generate nuclear wastes should not be forced to host a HLNW repository.
- Most also believed that building a single national repository was the least fair approach, compared with building two national repositories, or storing wastes in more than one region or state or at each reactor site.
- Two-thirds to three-quarters of respondents expressed serious distrust of DOE; they believed the agency would not be forthright about accidents or problems with the HLNW management program.

These surveys demonstrated unequivocally that most Nevada residents are strongly opposed to the repository program. A 1989 survey revealed that 70% would vote against the project, and nearly 75% believed the state should persist in fighting the repository even if this fight meant giving up benefits offered by the federal government. Follow-up surveys in 1990, 1991, and 1993 confirmed that very high levels of opposition and distrust persisted; the percent of Nevadans who would vote against a Yucca Mountain repository has remained in the 70% range, and opposition has continued to outstrip support by margins of three or four to one. Nevadans believe strongly that the state should be able to decide whether to accept the repository, and that state and local officials should be genuinely involved in decisions about Yucca Mountain. Finally, Nevadans expressed concerns about potential stigma impacts on tourist and visitor industries, which dominate the state's economy and revenue base.

To understand potential stigma effects, several studies were done outside Nevada to assess whether the mere existence of a nuclear repository within a 100-mile radius would reduce a community's attractiveness as a place to attend a convention, take a vacation, raise a family, retire, or locate a new business. Surveys were conducted among the general population nationwide, as well as among special groups, such as convention planners and real estate experts. The results clearly show that the Yucca Mountain repository could have a negative stigma effect on Nevada and particularly on Las Vegas, the major metropolitan area in the state. These findings include the following:

- A large majority of respondents said a repository would reduce the desirability of a community for raising a family.
- A majority said a repository would deter them from visiting for a vacation or attending a convention.
- Job seekers could be deterred from seeking work in Las Vegas because of the repository and because of reduced job opportunities resulting from the repository's presence.
- At least 30% of convention planners surveyed said they would reduce their rating of Las Vegas as a meeting site if a repository was located nearby. If the facility should experience accidents or incidents that are given extensive media coverage, 75% of planners would reduce their rating, and nearly 50% said they would no longer consider Las Vegas an acceptable convention site.
- Real estate executives believed that the existence of a repository within 100 miles of a community would detract from its suitability as a location for administrative offices, business and professional services, and businesses to serve the hospitality industry.

Socioeconomic surveys such as these are limited in their ability to predict actual future behavior, especially in relation to a unique facility with which no one has had any experience. Therefore, another set of studies based on the concept of *environmental imagery* was designed to test three propositions: (1) that people associate images with different environments and places, and these images can affect their behavior with regard to those places; (2) that a HLNW repository evokes extremely negative images; and (3) that negative images associated with the Yucca Mountain repository will extend to Nevada and Las Vegas. If true, these propositions suggest a mechanism by which the Yucca Mountain repository could generate significant social and economic stigma effects.

In these environmental imagery surveys respondents were asked to state the first thought or image that came to mind when presented with stimulus phrases such as "Las Vegas," "Nevada," and "nuclear waste repository." The findings, which supported all three propositions, indicate that

- the more positive an image a city or state elicited, the more likely it was to be preferred over other places for visiting, raising a family, retiring, or locating a business. Follow-up interviews with respondents 16 to 18 months later revealed that their earlier image scores significantly improved the ability to predict their vacation behavior during the intervening time.
- those who associated Nevada with things nuclear gave Nevada lower imagery and preference ratings than those who did not.

- the most common words associated with a nuclear waste repository were extremely negative, evoking images of danger and death. These verbal reactions suggest a profound aversion to nuclear wastes, comprised of feelings of dread, revulsion, and anger—the raw materials of stigmatization and political opposition.

The negative images associated with nuclear wastes come from the association with nuclear weapons, from events like the Three Mile Island reactor accident in 1979, and from revelations about radiation releases and leaking nuclear wastes at DOE's nuclear weapons facilities (such as the one at Hanford, Washington). When events like these—and the public's reaction to them—receive widespread media coverage, the stage is set for a phenomenon known as the *social amplification of risk*. For however many people experience the event directly, many more read and hear about it, and their assessment of the risks involved are influenced by the volume of news coverage, by the dramatization that often occurs, and by the degree of dispute about what happened and why. A highly amplified event can prompt people to apply extremely negative imagery to a place perceived as undesirable or hazardous.

One of the most notable byproducts of this amplification process, and the negative images it evokes, is its effect on the public's impressions about the ability of experts to manage and control technological hazards. Risk analysts have argued that public acceptance of a risk depends more on their trust and confidence in the risk managers than on technical estimates of how likely risks are to occur. This fact is often not appreciated by scientific, governmental, and industrial managers of nuclear technologies. These managers tend to place great reliance on technically-based risk assessments and are more confident than the public that they can safely handle and store nuclear wastes. Such managers often have difficulty acknowledging that the widening gap between the public's perception of risks and their own has created a crisis of confidence that will not be easily surmounted. While trust can be quickly lost, it is slowly, if ever, regained.

Governmental and Institutional Failures

We have seen that public distrust of the institutions managing nuclear wastes has contributed significantly to the problems experienced by the Yucca Mountain repository program. The continuing inability of the federal government and its agencies to forge workable relationships with state and local authorities has had an equally negative impact, making it almost impossible to resolve conflicts arising from political, administrative, and management issues. In fact, by the early 1980s, the relationship between DOE and many state governments had deteriorated to the point of open hostility. DOE's attempts to locate a repository site—conducted largely without any effort at local participation or consultation—had managed to provoke two-thirds of the states into banning site exploration within their borders.

NWPA was intended to resolve some of these hostilities, although its critics did not hold out much hope that it would do so. One analysis by an advisory panel to the U.S. Secretary of Energy identified as a major disadvantage the DOE's continuing inability to form "stable and credible relationships with states, tribes, interest groups and the public" (DOE, 1984, p. IX-3).

Although NWPA called for consultation and cooperation with potential host states and Indian tribes in the site selection process, all the real power was placed in the hands of Congress and DOE. All too frequently, DOE interpreted consultation to mean simply giving notice of an impending action, or worse, to engage in a promotional public relations effort. Moreover, DOE did not see any need to engage in meaningful negotiations with potential host states and could largely ignore these states' objections as long as it retained support in Congress, the ultimate arbiter. Even the state veto over site selection granted under NWPA was severely limited and unlikely to be effective against the greater power of Congress.

Thus, NWPA failed to include provisions to improve institutional relationships between state and federal authorities that would expedite repository siting. NWPA gave states and Indian tribes no final authority on any issue of substance and assigned responsibility for implementation to an agency that few people trusted to develop a fair siting process or to resist politically-expedient solutions.

The subsequent abandonment of NWPA's equity provisions only made the situation worse. Since Congress passed NWPAA in 1987, DOE has tried various means to force Nevada's compliance to siting a repository at Yucca Mountain—for example, by alternately threatening funding cuts and offering rewards designed to encourage community acceptance, while at the same time attempting to fragment opposition within the state. This strategy has utterly failed to obtain Nevada's cooperation with or acceptance of the repository program.

What DOE has not tried is to create new institutional roles for the states and the public in the repository program. Despite continuing problems generated by state, local, and public opposition to how the repository program is developed, DOE seems to recoil from suggestions that it pause to reconsider and revamp the existing unworkable institutional relationships. This restructuring would mean retreating from a forced siting approach at Yucca Mountain.

International Experiences

Like the United States, other countries have opted for deep underground burial of HLNW. They are grappling with the same issues and problems that the United States program has faced, but in most cases these countries are handling the problems differently and more effectively; other countries are adapting more rapidly and with more flexibility to HLNW challenges than the United States.

Sweden, which has had the most notable success, opened an interim spent-fuel storage facility in 1985 and an underground facility for low- and medium-level wastes in 1988. Sweden has also developed a waste transportation system and is making steady progress in finding a site for permanent disposal of HLNW. This disposal has been accomplished in a country noted for strong environmentalism, a tradition of local control and community veto rights, and a well developed anti-nuclear movement. And it has been done with only a small fraction of the people and money that the United States has devoted to its program.

Several factors account for Sweden's progress:

- A sincere commitment to two-way communication with the public; extraordinary efforts have been made to elicit public debate and feedback.
- Putting safety first and ensuring high quality technical scrutiny of the program by experts from around the world.
- Maintaining flexibility with the existence of an interim storage capability that buys time to conduct careful site studies for the permanent repository and to encourage community dialogue.
- Relying on long-lived engineered barriers, in addition to geologic barriers, to isolate HLNW, so as to err on the side of safety and minimize the impact of unforeseen events.

The Swedish approach reflects a determination to accommodate rather than override local concerns and to fashion a national consensus that will pave the way for a successful repository siting program. Other countries have taken a similar track. The Canadian program, for example, has adopted a policy that no effort will be made to site a repository until, or unless, the public accepts the disposal concept, and no attempt will be made to force a repository on a community. About 3% of the Canadian waste disposal budget has been earmarked for consultation and negotiation with the public and potential host communities.

Even France, which in the past pursued a hardball strategy and strong-arm tactics in dealing with anti-nuclear opponents, has come to accept that a waste disposal program must find some measure of public acceptance. This transformation occurred despite a bureaucratic culture that has viewed HLNW management as a strictly technical matter, not a political or social issue. The French have also opted to build two underground test labs into which wastes will be experimentally placed to study the response of the geologic structures. Eventually one lab will be developed into a full-fledged repository. This approach gives the program added flexibility; if one site is found to be unsuitable, a second will already be up and running. If both are acceptable, the French will have a choice of sites for the permanent repository.

In the United Kingdom stiff and widespread resistance to HLNW disposal caused the government to abandon its repository program in favor of interim storage in 1981, thereby handing the problem to future generations. Recently, however, the British program has shown signs of a more open, consultative process than the previous decide-announce-and-defend approach with which it tried to override public debate.

The German program is the one most similar to that of the United States. It, too, is in trouble, beset by a familiar litany of problems: public opposition, intergovernmental conflicts, legal challenges, and political delays. Because Germany's decentralized political structure permits considerable local influence over national policies, these battles have halted development of a German waste disposal complex several times. Germany has yet to adapt its program and institutions to cope with these political realities.

The European experience demonstrates that the United States is becoming increasingly isolated in its attempt to override rather than address state and local concerns. Most countries have determined that success depends on developing approaches that are socially acceptable as well as technically sound, collaborative rather than pre-emptive, and predicated on persuasion and negotiation rather than on coercion. These international efforts underscore the puzzling rigidity and lack of flexibility of the United States program and thus its vulnerability to failure. For example, most other countries (except Germany) are not in a hurry to bury HLNW. Instead of setting arbitrary deadlines for building permanent repositories, they have developed interim storage facilities that buy time so they can carefully examine potential repository sites while at the same time working to achieve social consensus and public support. The United States is, in fact, the only country in which the federal government is required by law to accept wastes for disposal by a set date. The United States is also the only country that lacks a long-term interim-storage program.

Except Germany, the United States is the only country that plans to characterize a single site, greatly increasing program vulnerability. Moreover, the United States is alone in its intention not to rely on engineered barriers to isolate wastes; instead, the geological structure of the selected site will have to do the job—a strategy that increases the risks associated with characterizing only one site that was selected for political rather than geotechnical reasons. Finally, the regulatory framework of the United States program is far more rigid than that of most other countries.

If the United States wishes to turn around the failing fortunes of its national HLNW management effort, it would do well to learn from the experience of other countries during the past decade. In particular, the United States must recognize and accept the following principles:

- Social issues must be put on an equal footing with technical matters.
- Coercive approaches to siting a repository don't work; they carry a high political price and provoke increasing public opposition.
- Public concerns are deep-seated and unlikely to be lessened by purely technical assessments, reassurances, or high-handed recourse to expert opinion.
- A highly integrated and adaptive management structure is preferred and necessary.
- The process of siting and building a HLNW repository will take a long time and require continual mid-course corrections; everything possible should be done to maintain flexibility within the program.

Recommendations

The HLNW program in the United States is failing badly, beset by technical difficulties, poor management, scientific uncertainties, cost overruns, equivocal political support, state opposition, and profound public mistrust and antipathy. It is doubtful the existing program can overcome these obstacles. A new approach is urgently needed. The following recommendations outline some crucial elements of such an approach.

Reevaluate the Commitment to Underground Geologic Disposal

Congress should place a moratorium on the current program. Further research should be done on technical problems, including study of the comparative advantages and disadvantages of different geological structures and engineered barriers to isolate HLNW from the environment. Most importantly, the federal government must make a genuine effort to gain public acceptance and political support for the program.

Use Interim Storage Facilities

Above ground storage in dry casks at reactor sites or a centralized monitored retrievable storage (MRS) facility could be used to store wastes for 100 years or more. This would allow the program to respond to the essential technical and socioeconomic problems rather than being driven by the current arbitrary schedule.

Evaluate More than One Site

Every effort must be made to find several states and communities willing to be considered as the location of an interim or permanent storage facility. It is crucial to keep several options open until very late in the selection process, because the repository is a first-of-its-kind facility with a great many associated uncertainties and a well-demonstrated ability to evoke intense public and political opposition.

Employ a Voluntary Site Selection Process

Procedures to select Yucca Mountain for site characterization have been a major source of conflict and have evoked fierce public and political opposition. To avoid such conflicts in the future, Congress should mandate that no community will be forced to accept a repository against its will and that potential host communities should be encouraged and permitted to play a genuine and active role in the planning, design, and evaluation of the repository. A voluntary process requires not only public participation, but also an agreed-on procedure (e.g., public referendum with a two-thirds plurality) for determining whether to accept the facility. Such a voluntary process must be given every chance to work.

Negotiate Agreements and Compensation Packages

A voluntary siting program must offer sufficient benefits to potential host communities and regions so that their residents feel their situation has improved over the status quo.

Acknowledge and Accept the Legitimacy of Public Concerns

A repository program has social and economic dimensions that will seriously affect the quality of life in neighboring communities. Most notably, such a project has the potential to stigmatize these communities, making them less attractive to residents, visitors, businesses, and in-migrants. The full range of socioeconomic impacts should be addressed as part of a negotiated siting process.

Guarantee Stringent Safety Standards

Public acceptance of the repository program requires assurances that public safety will be a priority. The federal government must negotiate contingent agreements with any community or region that agrees to host a repository and specify what actions will be taken should there be accidents or unforeseen events, interruptions of service, changes in standards, or the emergence of new scientific information about risks or impacts.

Restore Credibility to the Waste Disposal Program

The history of the national HLNW management program to date underscores one glaringly obvious point: DOE has failed in its management role and is incapable of overseeing such an extraordinarily complex and uncertain program. Congress should establish a new agency or organization to manage the civilian HLNW program, which should be separated from the military program.

No matter which agency is assigned the job, however, a radical new management approach is needed—one committed to implementing the recommendations listed above and doing so in an open, consultative, and cooperative manner. Given large inherent uncertainties with the repository program and the likelihood that it will encounter unforeseen surprises, the management approach must emphasize flexibility and adaptability.

In addition, the new management approach requires reliability (do the job without error), durability (survive for the length of time needed to complete the job), stability (continue performing the required tasks despite external changes), capacity (handle the required volume of wastes) and integrative skills (consider system-wide consequences of decisions).

Above all, we must remember that building a permanent, underground HLNW repository is essentially an experiment—one whose full social and economic dimensions are uncertain and unpredictable and likely to remain so. Citizens must decide how—and perhaps even whether—to proceed with this experiment in the face of all the unknowns and potential risks it presents.

2

Problems with High-Level Nuclear Waste and a Little History

The United States is facing one of the greatest scientific and technological challenges in its history: developing a national system for managing and storing high-level nuclear waste (HLNW). This system must operate safely for 10,000 years with extraordinarily few mishaps and cope with the uncertainties and surprises the future undoubtedly will bring.

Spent fuel rods from nuclear power plants, the primary source of HLNW, are currently stored in water pools next to reactors. Nuclear weapons production also generates HLNW. Their wastes are not only highly radioactive but also very hot, making them even more difficult to handle. Although other items used in nuclear installations, such as clothing, gloves, and equipment, are also contaminated, they present less of a hazard and are classified as transuranic or low-level wastes. But the HLNW remain dangerously radioactive for thousands of years; eventually they must be sequestered in a permanent facility. The United States, like several other countries, plans to bury these HLNW deep underground in a repository isolated from the surrounding environment by a massive barrier of stable, impermeable rock.

For more than 20 years, the United States federal government has been searching for a suitable geologic site for the underground repository and, at present, only one site is being considered: Yucca Mountain, Nevada. This siting process, and the policies on which it is based, is failing badly. It is threatened by serious technical and political obstacles such as:

- Doubts about the Department of Energy's (DOE) ability to effectively manage the waste program. This doubt stems from repeated schedule delays, escalating cost estimates, and DOE's past failures in handling HLNW from defense facilities.
- Scientific uncertainties about the suitability and safety of Yucca Mountain as a site for a HLNW repository.
- Wavering support from Congress and nuclear power utilities.

- Profound public distrust of the project and its managers.
- Sustained and creative opposition campaigns from the state of Nevada, bolstered by growing evidence that the repository may adversely affect a state economy predominantly dependent on tourism and gaming.

Before discussing these issues in detail, it is useful to examine the nature and genesis of the HLNW problem and the choice of geologic disposal as the preferred solution.

The High-Level Nuclear Waste Dilemma

Nuclear fuel consists of uranium oxide pellets inside zirconium-clad fuel rods. The pellets contain two forms of uranium; about 3–4% is fissionable U-235, and the remainder is nonfissionable U-238. Energy is generated by fissioning (splitting) the U-235, and when it is depleted the fuel rod is spent. However, while spent fuel rods are no longer useful, they are far from harmless. They have been irradiated with radioactive byproducts that vary greatly in level and type of radiation they emit (e.g., alpha particles, beta particles, gamma rays) and in their half-life (the amount of time it takes for their radioactivity to decay to half its original level). Some byproducts, such as strontium-90 and cesium-137, are dangerously radioactive for decades, posing potent short-term threats to human health. Others, such as technetium-99 and plutonium-239, remain dangerous for thousands of years. Consequently, spent fuel rods must be isolated from human contact for at least 10,000 years and possibly up to 100,000 years (DOE, 1990; Holdren, 1992; Lenssen, 1991).

At present, no method for isolating spent fuel rods exists; they are accumulating at 72 sites with a total of 110 reactors around the country. Each reactor discharges an average of 30 metric tons of spent fuel a year. And to date more than 20,000 metric tons of HLNW have been stored in water pools at reactor sites—an amount projected to double by the year 2000 (DOE, 1990; Holdren, 1992; Lenssen, 1991). The pools provide immediate cooling and radiation protection after the fuel is removed from the reactor core, but the pools are intended as a temporary solution.

After about 5 years, these HLNW are cool enough to be removed from the pools, although these wastes remain highly radioactive. But the question of where to put these wastes next is still unresolved. A new technology, known as dry-cask storage, was developed recently and permits above ground storage of HLNW at reactor sites for a century or more. But neither the water pools nor the dry casks—nor, for that matter, the utility companies that run reactors—will endure for the many millennia that HLNW remain dangerously radioactive, hence concern with developing a permanent system for handling and storing HLNW in the long run.

It was once thought that reprocessing (i.e., recovering uranium and plutonium from spent fuel rods) might provide part of the answer. Uranium could be recycled into new fuel rods, while plutonium could be used for nuclear weapons or plutonium-fueled reactors. In the 1960s reprocessing of civilian nuclear fuel started at West Valley, New York, but it never achieved viability as a waste management technique. (Unfortunately in the process West Valley became a major waste site, and it is now subject to a large-scale clean-up effort.) Proposed reprocessing facilities at Morris, Illinois, and Barnwell, South Carolina, were also abandoned in the mid-1970s before they were put into operation due to technical, financial, and political problems. Revelations during the 1980s of difficulties and contamination problems experienced by reprocessing at DOE weapons production facilities underscored the demanding technical problems and countless opportunities for failure in waste management. And at the national policy level, concern that reprocessed plutonium might contribute to proliferation of nuclear weapons worldwide grew (Carter, 1987).

The failure of reprocessing as an option for interim waste management increased the sense of urgency to find an alternative. Moreover, because of the long-lived radioactivity of spent fuel rods, many citizen and environmental groups made demands for a permanent disposal method. Antinuclear groups were especially adamant in calling for such a method, arguing that society should not sanction any technology that produces unmanageable waste.

In 1976 the California legislature forged a direct link between permanent disposal and continued development of nuclear power by enacting a moratorium on building new reactors until a federally-approved permanent disposal method was devised. This bill was introduced by Assemblyman Charles Warren to preempt a proposed state referendum (Proposition 15) that would have placed even stronger constraints on nuclear power development. The bill gave the state authority to consider the economic implications of nuclear waste disposal in deciding whether to license nuclear reactors. Specifically, California took the position that lack of an established federal waste disposal program, and uncertainties about its future costs, posed too great an economic risk to electric utilities' rate payers. The Supreme Court upheld the bill on these grounds of potential economic risk. Similar moratorium laws were subsequently enacted in Oregon, Wisconsin, Maine, Montana, and Connecticut.

Geologic Disposal

Since the 1950s, nuclear industry scientists and the federal government have regarded geological disposal as a safe method for long-term storage of HLNW. The concept, also known as *reverse mining*, involves entombing wastes deep underground in geological structures (e.g., salt beds, volcanic tuff, basalt, granite) known for their stability, impermeability, and isolation from the outside environment. The current plan, still in the design stage, calls for a series of tunnels to be drilled deep into a geologic structure, with each tunnel containing bore holes into which waste containers would be placed. The containers, made of corrosion-resistant stainless steel and designed to shield radiation from the environment for 1000 years, would provide an extra, engineered barrier to augment the geologic barrier. After the containers are placed into individual bore holes, the holes would be sealed with a liner and closed at the surface.

Limited retrievability of the wastes would be possible for about 50 years while containers were still being buried, but once the facility was filled to capacity, the entire repository would be backfilled and permanently sealed to make the wastes inaccessible.

Geologic repositories have also been designated as the final resting place for HLNW generated in nuclear weapons manufacturing. Much of this waste is currently stored at Hanford, Washington, in underground storage tanks that have been plagued by leaks, including one from which up to 800,000 gallons of radioactive water escaped into the surrounding soil (GAO, 1991). Moreover, cyanide compounds put into the tanks to reduce waste volume created risk of chemical explosion (GAO, 1990). A considerable amount of weapons wastes are also stored at Savannah River, South Carolina, and there are plans to stabilize the more dangerous of these wastes by *vitrification*, a process of mixing wastes into a glassy liquid and then solidifying them, and enclosing the wastes in stainless steel containers.

In the United States geologic disposal became the officially-sanctioned option for HLNW disposal with passage of the Nuclear Waste Policy Act (NWPA) in 1982. NWPA committed the federal government to building at least two underground repositories—one in the west, where some site investigation had already been done, and one in the east, where most of the nuclear wastes are generated. Congress decided to pursue this option after several federal studies considered various alternatives, including deep seabed disposal and launching the wastes into space (DOE, 1978). However, in 1987 passage of the Nuclear Waste Policy Act Amendments (NWPAA) aborted the site selection process and limited the repository program to a single location: Yucca Mountain, Nevada, approximately 90 miles north of Las Vegas. The current plan calls for the repository to be dug into a geological formation known as welded tuff, about 1000 feet beneath the surface of the mountain.

However, the repository cannot be built at Yucca Mountain until DOE characterizes the site—that is, determines whether the geologic structure is sufficiently stable and impermeable to contain radiation from HLNW for the required 10,000 years. The data collected during this exercise will be used in an application for a construction license to the Nuclear Regulatory Commission (NRC), which determines whether scientific evidence indicates the repository will meet radiation-release standards to be established by the Environmental Protection Agency (EPA).

Although this process is still in the early stages, site characterization at Yucca Mountain has become mired in a political, financial, legal and public relations quagmire from which it appears unlikely to be extricated. The schedule has already slipped badly. NWPA originally called for the repository to open in 1998, but DOE's continuing problems with the program have twice pushed back that date, first to 2003 (DOE, 1987) and currently to 2010 (DOE, 1989). The latest target assumes that DOE can meet its current schedule, and, more importantly, that Yucca Mountain proves to be a viable site. Both assumptions are being openly questioned, and further delays seem inevitable, especially because none of the scientific and political problems that have plagued the process so far show any sign of abating.

Failings of the Repository Program

Technical and Scientific Uncertainties

There is considerable scientific uncertainty as to whether the Yucca Mountain site can safely isolate HLNW from the surrounding environment and communities for 10,000 years. DOE is expected to demonstrate that nearby populations will receive only minimal radiation exposure, but determining just how much radiation might escape over such a long period challenges the abilities of physical and social scientists to anticipate and plan for future contingencies.

Concerns are being raised that catastrophic geologic events or human intervention could threaten the security of a repository at Yucca Mountain. For example, one of DOE's scientists, Jerry Szymanski, has proposed a theory that Yucca Mountain is subject to upwelling of groundwater—water forced up through the mountain by a major seismic event, such as an earthquake. He argues that there is geological evidence this event has occurred in the past. If it were to occur with a repository in place, water would presumably flood the waste storage area, possibly causing the canisters to break and release radiation to the environment. A plume of contaminated water released high on the mountain could have widespread and disastrous consequences (Broad, 1990).

There is not any scientific consensus on this scenario, and its plausibility has been disputed by scientists in government agencies and academic institutions. A DOE-funded review of the theory by the National Academy of Sciences concluded that such a catastrophic event was unlikely (NRC/NAS, 1992). However, subsequent research sponsored by the state of Nevada supports the possibility that this event could happen (Shrader-Frechette, 1993).

The licensability of Yucca Mountain is also threatened by the possibility that strong earthquakes might occur in the surrounding region. DOE and the state of Nevada have publicly disagreed about whether the fault system surrounding Yucca Mountain will give rise to earthquakes that would breach the containment system. This question was given added prominence on June 29, 1992, when a magnitude 5.6 quake occurred at Little Skull Mountain, just 12 miles from the proposed repository site. Reaction to this incident provides yet another demonstration of the gulf between proponents and opponents of the Yucca Mountain project. While DOE officials argued that the event demonstrated the geologic structure was robust enough to warrant a repository, Robert Loux, executive director of the Nevada Nuclear Waste Project Office (NWPO), described it as a "wake-up call" that "confirms that the whole of southern Nevada is a young, active, geologic environment" (Rogers, 1992). This observation was reinforced by a magnitude 6.0 quake on May 17, 1993, about 100 miles west of Yucca Mountain and 35 miles southeast of Bishop, California. The Ghost Dance Fault, which cuts directly through Yucca Mountain from south to north, is of particular interest. As government geologists explained in May 1993, they do not know when this fault was last active or whether it is in a zone connected to other faults (Rogers, 1993).

Scientists have also found evidence that two volcanoes located within 27 miles of the Yucca Mountain site erupted as recently as 5000 years ago, which suggests that volcanoes could erupt again within the 10,000 years that repository wastes will remain dangerously radioactive (Shetterly, 1988).

Another potential problem is the release of radioactive carbon-14. There is consensus among scientists that some carbon-14 will escape over the repository's lifetime. In fact, a relatively large fraction of the carbon-14 present when the spent fuel rods are emplaced will eventually reach the environment (van Konynenburg, 1991) because carbon-14, which has a half-life of 5730 years, will still be present in significant quantities when the waste canisters exceed their effective life span of 300–1000 years. Once a canister is breached, the carbon-14 will readily transform into its gaseous state and escape to the surface before decaying to harmless levels.

Finally, future human intrusion, either accidental or deliberate, has been suggested as a potential threat to the integrity of the repository. Intentional mining of HLNW could conceivably occur if future generations come to regard the wastes as valuable, either because an effective means of reprocessing is developed or because some other use is discovered. Accidental intrusion could occur if future generations forget that the wastes are buried at Yucca Mountain. It is certainly open to question whether human memory of the repository site can survive 10,000 years of cultural, climatic, and geologic changes that will probably result in settlement patterns, economies, and languages that may bear little resemblance to those of today's society (Erikson, 1994; Erikson, Colglazier, & White, 1994).

These risk scenarios suggest that a repository at Yucca Mountain might not isolate radiation to the degree required by EPA standards. In fact, DOE may find it impossible to present sufficient evidence to demonstrate that Yucca Mountain is a suitable repository site. In 1990 the National Research Council's Board on Radioactive Waste Management identified several sources of scientific uncertainty that prevent DOE from proving various radiation release events are sufficiently unlikely, including the inability of scientists to collect all relevant data, the evolving nature of geohydrologic computer modeling, and the extremely long period during which risks will persist. The board said these uncertainties make it impossible to estimate with any precision the likelihood of many of the release events that would be covered by EPA standards.

Congressional response to this difficulty was to pass the 1992 Energy Policy Act, which mandated a less stringent method of calculating radiation risks for the Yucca Mountain site than EPA planned to implement (see chapter 3). This new approach, which applies only to Yucca Mountain, will make it easier for DOE to obtain a construction license from NRC. The Energy Policy Act also attempted to legislate away the problem of potential human intrusion by requiring DOE to act as a perpetual custodian over the repository. It is not clear how a legislative edict like this can, in any practical sense, ensure 10,000 years of continuous stewardship. No theoretical or historical evidence shows that any societal or government institution can endure for such a long time. DOE, for example, was created in 1978 and is the successor agency of the Atomic Energy Commission (which existed for less than 40 years) and the Energy Research and Development Agency (which existed for 4 years).

While the 1992 Energy Policy Act eases some of the obstacles DOE faces in licensing Yucca Mountain, it does not ensure approval of a repository. If the scenarios raised by Szymanski and others turn out to be credible, they will certainly preclude the NRC from issuing a license for Yucca Mountain, regardless of standards EPA eventually promulgates. In the meantime, DOE and the state of Nevada continue to disagree on the site's technical suitability. DOE contends that its data analyses support a more positive view, or at least that more information is needed to see if the more positive view is valid. Perhaps the only thing that is abundantly clear is that many important scientific and technical uncertainties remain unresolved—so much so that different groups of analysts can look at the same data and legitimately reach exactly opposite conclusions.

Political Opposition

The state of Nevada and a majority of its residents are strongly opposed to the proposed Yucca Mountain repository. As a result, state officials have initiated numerous legal actions against the DOE site selection program. In 1985 and 1986 law suits were filed against the U.S. Secretary of Energy to force DOE to comply with the letter and intent of the 1982 NWPA. Then, in 1987, after NWPAA designated Yucca Mountain as the only potential site to be considered, political opposition in the state escalated sharply. In 1989 the state legislature passed two resolutions in opposition to the repository and enacted Assembly Bill 222, which made it illegal to dispose of HLNW in Nevada (Swainston, 1991). Subsequently, the state attempted to deny DOE environmental permits for its site characterization effort. Although the state did not win these legal skirmishes, a considerable amount of time was needed to resolve the issues.

DOE claimed these actions forced it to divert attention and resources from Yucca Mountain site characterization work, causing serious program delays. The state contended that it was not responsible for the delays, citing a GAO report that concluded DOE was so slow in preparing for the scientific work that its people would not have been in the field even if Nevada had cooperated with the siting effort (GAO, 1992). According to Governor Miller, the report provided evidence that DOE had used Nevada "as a scapegoat—one more scapegoat in a whole DOE herd—to hide DOE's own incompetence, mismanagement, and failure." Many observers, however, feel the governor was too modest; although DOE was not ready to go into the field in 1990, the state's actions did sap its resources, and site characterization is now farther behind than it would have been if the state had cooperated.

And there is every indication that Nevada has the ability and resolve to further delay site characterization. For example, state officials could conceivably deny DOE the permit needed to carry out site characterization. This denial was a DOE concern in 1991 and during the 1991–92 session of Congress when legislation was introduced to strip Nevada of its permitting authority. This provision, included in President Bush's National Energy Strategy, was passed by the full House and by the Senate Natural Resources Committee but did not make it into the final version of the Energy Policy Act. In a future case, the U.S. Secretary of Energy might need to file suit against a state agency claiming that the actions of state officials were capricious or arbitrary. This claim may or may not be difficult to prove, depending on how the application was filed, but in any event, DOE's work could be delayed until the case was adjudicated.

On the other hand, it is unclear whether the governor would be willing to allow state officials to exercise this authority. The hearings before the state engineer on DOE's water permit application revealed the risks of this strategy. It became clear that if the state engineer rejected the attorney general's argument about the repository's negative effect on the physical and economic health of the state it would be difficult for the attorney general to later argue such a case in a federal lawsuit.

Recent actions by Idaho in limiting and controlling shipments to the Idaho National Engineering Laboratory (INEL), the ability of public opinion to influence and delay DOE plans for the Waste Isolation Pilot Project (WIPP) in New Mexico, and the nationwide resistance to federal and state plans for constructing low-level radioactive waste facilities, all demonstrate how fragile schedules are when significant public and intergovernmental opposition takes place.

Even if DOE eventually succeeds in characterizing Yucca Mountain and the secretary finds the site suitable, the saga will not end. Under NWPA, Nevada can submit a notice of disapproval to Congress and make its case regarding disqualifying conditions to the NRC during the licensing phase. If the Szymanski theory about groundwater upwelling survives the data acquired during site characterization (in even a probabilistic sense), NRC will find it difficult to approve Yucca Mountain. If, on the other hand, NRC concludes that the evidence supports a finding of suitability, another round of lawsuits can certainly be expected.

Thus, the future portends a prolonged, litigious, messy conflict. Nevada may find either a legal principle or technical flaw that will effectively abort the plan to build a repository at Yucca Mountain. And even if DOE obtains a license for the site, this decision will likely be considerably delayed by legal maneuvering. It has been estimated that Nevada can effectively postpone the siting process for up to 20 years (Rhodes, 1990).

Program Delays and Erosion of Political Support

Technical uncertainties about the suitability of the Yucca Mountain site and Nevada's unrelenting political and legal opposition to a repository have raised widespread doubts about whether this site will prove viable (e.g., Ahearne, 1990; Flynn, Kasperson, Kunreuther, & Slovic, 1992). Regardless of whether these doubts are well founded, they may doom the Yucca Mountain option because of DOE's seeming inability to get beyond these difficulties, plus evidence of its management failures, have undermined the rather tentative support of its advocates within the nuclear industry and Congress. Without the backing of these players, DOE will be left without the means to prepare an adequate license application.

Repeated delays in the repository program have already caused a considerable erosion in DOE's political support, and there appears to be no end in sight to this particular problem. Indeed, Energy Secretary Hazel O'Leary recently admitted that the official 2010 target date for opening the facility will be impossible to meet (Salpukas, 1993). And GAO concurs. In a recent report GAO estimated that the project is anywhere from 5 to 13 years behind schedule, depending on future funding levels (GAO, 1993). These delays were caused by Nevada's opposition, by a realization of just how big a job site characterization is, and by DOE's lack of management ability (GAO, 1992; NWTRB, 1993).

The utilities are concerned by this situation. There are doubts that the federal government can start accepting waste in 1998, as originally stipulated in NWPA. In fact, in 1983, DOE signed contracts with the utilities to start accepting title to the wastes in 1998, and that date has never been changed, even though the official date for opening the repository has been delayed. The utilities' expectations persisted despite postponements in the repository program, because DOE maintained that a monitored retrievable storage (MRS) facility—an above ground parking garage for waste casks that could warehouse spent fuel rods temporarily until the repository becomes operational—would be available by 1998, thus allowing the utilities to begin transferring wastes from water pools at reactor sites. However, NWPAA cancelled plans for an MRS facility in Tennessee and set up a commission to examine the need for such a facility. Congress also established an Office of the Nuclear Waste Negotiator to bargain with volunteer communities for an MRS or a repository, but the post was not filled for more than 2 years. In 1992 the negotiating approach was described as a failure by then-Energy Secretary James Watkins because the process would not allow DOE to meet the 1998 goal. The various attempts by the federal government and the nuclear utilities have so far failed to find a state or Indian tribe willing to enter into an agreement.

Even if a potential MRS site is found soon, considerable lead time would be required to obtain a license and issue contracts. Perhaps more importantly, there remains a statutory obstacle to opening such a facility by 1998. Under NWPAA, an MRS can begin accepting waste only after a construction license has been issued for a permanent repository.

This coupling of the MRS and the repository was designed to facilitate the voluntary siting process, by reassuring potential MRS hosts that the MRS would not become a de facto permanent storage facility. Because Yucca Mountain was being considered for a permanent repository, the legislation prohibits location of an MRS in Nevada. This stipulation was intended to ensure that the Yucca Mountain site characterization studies would remain objective; there were concerns that if an MRS facility was built in the state, a permanent repository would inevitably follow, no matter what was found during site characterization at Yucca Mountain.

However, under DOE's current characterization plan, the earliest date for preparing a license application is 2001, admittedly an unattainable goal. In short, the HLNW program cannot meet the congressional schedule for removing wastes from the reactor sites (NWTRB, 1993). Of course, the program has missed every other scheduled milestone too. This one, however, is more important to DOE and Congress because the nuclear industry is a powerful stakeholder group and possesses significant influence with the federal government (Jacob, 1990).

While DOE's failure to meet the prescribed timetable has compromised its reputation with the utilities and Congress, the Yucca Mountain repository program is not doomed. However, its political viability is threatened by the combination of these delays and the scientific uncertainties about the suitability of the site. If, after another decade, DOE is unable to produce evidence demonstrating the suitability of Yucca Mountain, the United States will be back to square one in its effort to locate a permanent storage place for HLNW. A similar situation will result if Nevada succeeds in blocking construction of a repository. Either way, the utilities will be left without a means of permanent disposal—a disturbing prospect that leaves many industry representatives seeking an alternative policy. In 1990 former NRC Chairman John Ahearne recommended abandoning the repository effort and switching to a waste-management strategy based on above ground storage facilities that could be located regardless of geologic conditions. In fact, an increasing number of utilities are assuming that Yucca Mountain will not be licensed in time to meet their interim waste storage needs and are building dry-cask storage facilities at reactor sites.

The utilities are also concerned about the financial implications of continuing the current policy. Under NWPA utilities are assessed a charge of 1 mil per kilowatt hour on all nuclear-generated electricity to support a Nuclear Waste Fund that pays for the repository program. Consequently, they have a direct stake in seeing that funds are wisely spent. The utilities have reacted skeptically to DOE's latest cost projections for characterizing Yucca Mountain. In January 1993 Carl Gertz, Director of DOE's Yucca Mountain Project Office, estimated that an average of $600 million must be spent annually over the next 7 years—more than $4 billion in total—to prepare the license application (NWTRB, 1993). This is in addition to the $2.4 billion DOE has already spent. Thus, the total bill would exceed $6 billion—a far cry from the original $60 million estimate to characterize a single repository site presented to Congress when it crafted NWPA in 1982. However, it is unlikely Congress will authorize increasing DOE's funding for site characterization studies to $600 million a year, double its current annual budget for the project, given restrictions on federal spending now in effect.

Proceeding with Yucca Mountain is a tremendous gamble for the utilities, given the scientific, financial, legal, and political uncertainties dogging the project, and their continued support for DOE is far from assured.

One can argue that the current policy for managing HLNW is failing on several grounds. A growing minority of scientists seriously question whether Yucca Mountain can safely contain HLNW over the long term. Residents of the proposed host state are even more skeptical, and their concerns have translated into potent political opposition that represents another serious roadblock to building a repository at Yucca Mountain.

These technical and political uncertainties have fostered a widespread perception among Congress and the nuclear utilities that siting a repository is beyond the ability of DOE—or any other federal agency for that matter. This perception is fueled by repeated delays and cost overruns. In 1982, the total time from program onset to construction of the first repository was projected at 15 years. Already more than a decade into the program, DOE does not expect to begin receiving wastes for another 17 years and few observers, including GAO, believe this estimate.

These failings have compromised support for the program in Congress and among the utilities. Few observers expect that DOE can convince Congress to double the annual appropriations for site characterization studies, which will be needed to complete a license application by 2001. (This application is a major reason why DOE would like to have direct access to the monies in the Nuclear Waste Fund, rather than having them administered and paid out as part of the federal budget.)

There are signs that even DOE thinks the current program is in trouble. Despite earlier assurances that they could meet all safety and environmental standards, DOE has been pushing for a relaxation of these requirements. The change in radiation standards introduced in the 1992 Energy Policy Act was a tacit admission that Yucca Mountain would not meet the regulatory requirements EPA intended to promulgate, and indicated that the site may have trouble meeting any reasonable set of standards. And, in December 1992, then Energy Secretary James Watkins, on the eve of leaving office, acknowledged the difficulty of meeting the 1998 deadline and advocated that Congress return to the utilities some of the money contributed to the Nuclear Waste Fund for the repository program so that the utilities could build dry-cask storage facilities.

DOE has underestimated or misread everything about the repository program—the schedule, the costs, the difficulty of the task, the scientific uncertainties, and the political and social climate. All the doubts and uncertainties—and the consequent wavering in political support—have reduced its chance of obtaining full funding for the Yucca Mountain study, and thus its ability to prepare a successful and timely license application.

This sorry situation is far from the program envisaged when an optimistic Congress passed NWPA in 1982. NWPA, which contained a set of democratic principles and ethical rules to guide the selection of a repository site, appeared to be an innovative effort by the federal government to devise an acceptable process for developing not only the first repository but future ones as well. Unfortunately, more than a decade later many of these principles have been abandoned. Indeed, current policies and practices are driven by a misguided notion that obtaining the support and acceptance of stakeholders directly affected by the repository is an impediment to a successful siting program and that assertions of federal authority can suffice (Carter, 1993). Not surprisingly, this approach has only exacerbated public and political opposition and portends even greater difficulty for the future of Yucca Mountain or other future waste storage facilities.

3

Policies and Politics

The United States high-level nuclear waste (HLNW) management program has had a fractious and troubled history from the beginning. The 1982 Nuclear Waste Policy Act (NWPA) attempted to address some of these long-standing problems by establishing a set of democratic principles and ethical rules to be used in selecting a repository site, allocating its costs and benefits, and defining the relationship between the Department of Energy (DOE) and candidate states.

Unfortunately, the HLNW program has not lived up to these principles. In fact, nearly all have been compromised or repudiated by DOE and Congress with predictably perverse results. Instead of smoothing the way for repository development, the federal government's actions have intensified public and political opposition and severely jeopardized chances for successfully siting and building a geologic repository. Moreover, DOE's inability or unwillingness to establish an open and cooperative institutional relationship with states, Indian tribes, and local communities—a key objective of NWPA—virtually guarantees continued conflict.

To understand how policies went so badly off the tracks, it is necessary to first examine the provisions of NWPA in some detail.

Principles of the 1982 Nuclear Waste Policy Act

The ethical principles in NWPA were to protect public health and the environment, ensure an equitable and democratic decision-making site selection process, and gain support in Congress, even among members representing states in which a repository might be located. It was also hoped that a principled process would allay public concerns and reduce opposition from states eventually selected to host repositories.

Protect Public Health and the Environment

EPA was instructed to develop, by January 1984, standards for radiation exposure protection. EPA eventually determined that the repository could cause no more than 1000 excess cancer deaths over a facility life-time of 10,000 years. The Nuclear Regulatory Commission (NRC) was to apply EPA standards in setting its criteria for authorizing construction of the repository. To meet EPA standards, DOE would be required to pre-pare a lengthy license application identifying pathways by which radi-ation could be released. Only by documenting that radiation releases would be kept within EPA standards could DOE obtain approval to pro-ceed with construction.

Distributional and Outcome Equity

For a HLNW repository siting policy to have an equitable outcome, construction should result in an equal change in the welfare of all parties involved (whether individuals, corporations, or political jurisdictions) and this change should be beneficial. NWPA strove for outcome equity in three ways:

1. The financing provisions ensured that parties who benefit from a repository pay for its development. Consequently, the repository program would be funded not out of general tax revenues, but by nuclear waste generators and nuclear power beneficiaries. Therefore, the Nuclear Waste Fund [Section 302] was created, and its funds were to be collected by assessing a fee on nuclear-generated electric-ity. Utility companies, in turn, could collect this fee from ratepayers.

2. NWPA provided for monetary compensation to those who would live near the repository. Under Section 116(c)(3), state and local jurisdictions and affected Indian tribes were to be given payments in lieu of taxes that would otherwise be collected if the repository were a private venture subject to local, county, and state taxation. In addition, DOE was required to provide funds for impact assis-tance to affected levels of government to offset the repository's negative economic impacts [Section 116(c)(2)]. Thus, some benefits reaped by nuclear power users would be transferred to those who bore the costs of living near the facility. This transfer would, in turn, produce a more even distribution of the net benefits associated with the repository (Colglazier & Langum, 1988; Kasperson, Ratick, & Renn, 1988).

3. NWPA required two regional repositories. Prior to 1982, DOE was predisposed to build a repository only in the west, where it had al-ready done preliminary analyses of potential sites. However, almost all nuclear reactors are located east of the Mississippi River. Had DOE's original plans proceeded, the areas that would benefit most from a repository would escape the burdens associated with living

near it. Representative Morris Udall (D-Ariz.) offered a compromise that met the utilities' objective of expediting development of a repository, while also promoting geographic equity: DOE would build the first repository at a western site that had already undergone preliminary evaluation, but a second repository would also be required. The second repository would be located in a region of the country where DOE had not done previous siting studies—that is, in the midwest, northeast, or southeast.

The latter policy promoted distributional equity; it required eastern states to shoulder some of the burden associated with a repository. This policy also spread the burden across multiple states. Siting repositories in two states is more equitable than siting just one. In fact, placing a permanent repository at each reactor site would maximize equity, but this method would be inefficient as a permanent solution because repositories have tremendous economies of scale (e.g., for costs of studying the geology of potential sites, drilling out the repository, monitoring the site during pre- and post-closure). Fewer repositories also reduces the number of workers and residents potentially exposed to radiation, lowering overall health risk. This advantage in risk is, however, somewhat offset by the increased distances that wastes must be transported and the concentration of radiation hazard in one large, lethal package. Congress balanced equity with efficiency by choosing to build two repositories.

Procedural Equity

Equity entails not only a fair distribution of costs and benefits, but also fair procedures to determine the cost–benefit distribution—in other words, procedural equity. With NWPA, Congress attempted to ensure that fair procedures would be used in choosing candidate sites and bringing interested parties into the process. This procedure represented a departure from the situation that occurred in the 1970s, when DOE made efforts to site repositories in Lyons, Kansas, and Alpena, Michigan, without much interest in or regard for views of local communities or authorities. NWPA, in contrast, required DOE to open its site-selection decisions and scientific analyses for review by affected stakeholders, to notify governors when their state was being considered for a repository, and, once a site had been identified, to disseminate information on plans for studying the site. NWPA stated that affected states and Indian tribes were entitled to "timely and complete information regarding determinations or plans made with respect to site characterization, siting, development, design, licensing, construction, operation, regulation, or decommissioning" of repositories [Section 117(a)(1)].

In addition, during site characterization DOE was to make its scientific data and analyses available to affected states [Section 113(b)(3)]. Finally, the siting process was to be open not only to state and tribal officials, but also to the general public; NWPA required DOE to conduct public hearings at major decision points and make its plans for recommending sites and doing characterization studies known (Wiltshire, 1986).

NWPA made the siting procedure not only open but also objective. DOE was required to establish a set of technically-based guidelines for selecting and eliminating sites and adhere to these guidelines in narrowing the search to a final site. The U.S. Secretary of Energy was required to go through a series of decision steps and to report the results of each one, detailing which sites were still in the running and how they were selected. This multistep decision process appeared to many observers to be a good-faith attempt to identify the best possible sites for the two repositories (Colglazier & Langum, 1988; McKay & Swainston, 1989). Because each step was open to public scrutiny, DOE would have difficulty selecting a site on arbitrary or capricious grounds. And even if a politically favorable site were chosen in the first round, that site would have to prove out as the superior site to merit final selection because at least three sites would be characterized in the first round.

Another important aspect of procedural equity concerned the role of affected states and Indian tribes in site selection. DOE was required to keep them apprised of the siting process and to consult and cooperate with them before making key decisions [Section 117]. While this role did not give states and Indian tribes the authority to control the siting process, they did have the ability to make public scientific evidence that a proposed site was unsuitable. As Jacob (1990) pointed out, the consultation and cooperation provision lacked specifics as to how DOE should weigh state and tribal concerns; thus, DOE retained control over the siting process. In contrast, the initial proposal for consultation and concurrence (proposed by the Interagency Review Group and the National Governors Association) would have provided states with real decision-making power by giving them the authority to disapprove of DOE's decisions. If this led to an impasse, the parties would be subject to binding arbitration. Admittedly, the U.S. Secretary of Energy is not required to act on evidence of disqualifying conditions, but to ignore such evidence is senseless because the evidence would certainly be provided to NRC if DOE applied for a site license.

The right of a candidate state to identify disqualifying conditions was bolstered by two other provisions that increased the information available to the state. First, DOE was required to make available the scientific data collected during site characterization studies; in fact, progress reports were to go to the state governor every 6 months [Section 113(b)(3)]. Second, affected states and Indian tribes were given funding to conduct their own scientific studies. Specifically, NWPA allocated money from the Nuclear Waste Fund to candidate states to hire scientists to perform geologic, hydrologic, and socioeconomic studies [Section 116(c)]. These studies allowed candidate states to verify DOE's analyses (i.e., to act as a peer reviewer) and identify new problem areas needing study. By granting candidate states funding for independent studies, NWPA significantly advanced the evidentiary equity of the siting process.

Host states were also granted a say over the siting process by a NWPA provision allowing them to file a "notice of disapproval," which gives the governor and legislature of a state designated to host a repository the authority to veto site selection. This notice is a relatively weak measure (it can be overridden by a joint resolution of Congress with a majority in each chamber), but it went beyond what Congress had previously allowed in facility-siting situations. For example, the Atomic Energy Act did not allow state officials a formal voice over the licensing of commercial nuclear reactors.

The federal government's failure to establish open, consultative, institutional relationships with state and local authorities, relationships needed to transform these principles of procedural equity into reality, will be discussed later in this chapter.

Focus on Legitimacy

The provisions for public safety, outcome equity, and procedural equity were explicitly included in NWPA to create a more legitimate siting process and to make the eventual siting decision more acceptable to those directly affected. The importance of these objectives was stressed by then-Governor Dick Riley of South Carolina, who said:

> Of highest importance today is not only what is to be done, but also how we decide it is to be done. A process of decision making must be established that will allow us to have confidence in the results of that process. There will be remaining uncertainties no matter what the decisions are. Only confidence in the process which leads to those decisions will enable us, as a society, to live with those remaining uncertainties. (Colglazier, 1982, p. xi)

The political consensus NWPA attracted in its final form provided powerful evidence for the legitimacy of the policies it espoused. The Senate passed the bill by a voice vote under a unanimous consent agreement, while the House voted 256 to 32 in favor (Carter, 1987). Although the bill did not achieve unanimous consent under the final votes, the level of consensus was extraordinarily high for such a complex and consequential bill.

Further, NWPA was supported by a majority of senators and representatives from states identified as candidates for a repository. Most who had a direct interest in HLNW (e.g., Morris Udall of Arizona, James McClure of Idaho, Slade Gorton of Washington, Bennett Johnston of Louisiana, William Proxmire of Wisconsin) were able to influence the bill during debate, and the final version incorporated the ideas and objectives of the majority of the stakeholders who were then attentive to the HLNW issue: DOE, the utilities, the National Academy of Sciences, state officials, and some of the national environmental groups. Cook, Emel, and Kasperson (1992) described the process:

> The law emerged from just the kind of open, fluid, activist politics given birth in the reform period of two decades ago, with the nuclear industry, host states, environmental and anti-nuclear groups, and the Department of Energy (DOE) each vying to shape the legislation to their liking. At the center of the open, fragmented, high-velocity politics . . . stood members of Congress, attempting to finds areas of agreement . . . while also trying to meet constituency demands, career interests, and their own ideas about good public policy. (p. 52)

Thus, NWPA appeared to be a good-faith attempt to put together a siting process that could gain wide acceptance. The high level of consensus can be taken to indicate that the potentially affected parties, through their representatives in Congress, consented to take part in the siting procedure as specified in NWPA, which not only conferred legitimacy on the procedure but also implied that its outcome should be accepted.

NWPA undoubtedly contained some promising and innovative features, but policy analysts disagree on how successful NWPA was. Some observers viewed it as a relatively effective effort at achieving a political compromise that could reconcile the competing interests and objectives of diverse stakeholders. But Carter (1987, p. 195) described it as a "fragile consensus," and Jacob (1990, p. 95) asserts that it "did little to resolve concerns about the credibility, legitimacy, and financial condition of the nuclear establishment" and left "long standing conflicts" in place. The central thesis of Jacob's (1990) book is that NWPA constituted a major capitulation to the nuclear establishment, especially on issues such as participation by the host states, the lead agency and the amount of time that should be spent in evaluating alternative sites and technologies before making a final decision. Kraft (1992) argues that NWPA failed because of poor design and because of "state and public rejection of DOE proposals" and that NWPA "resulted in political gridlock" (p. 152). In any event, neither the principles nor the provisions of NWPA survived long.

Violation of Ethical Principles

By attempting to balance the country's need for a permanent HLNW repository with ethical principles designed to ensure a fair siting process, NWPA went a long way toward reassuring residents of the host state that the site finally chosen would be the best in that region of the country. However, many key ethical principles of NWPA were violated by subsequent actions of DOE and Congress. These violations, and DOE's implementation of the program, provoked concerted and sustained opposition nationwide and especially within Nevada.

Selecting Hanford over Richton Dome

An early violation occurred in February 1985 when then-Energy Secretary John Herrington tentatively earmarked three western sites for further in-depth study: Yucca Mountain, Nevada, Hanford, Washington, and Deaf Smith, Texas. This decision caused considerable surprise; it had been assumed that DOE would study sites that the preliminary assessments had identified as the most promising technically, but many observers believed that Hanford was not among the top three.

The expected acquiescence of the community near Hanford appeared to be a factor in its selection; the presence of a major nuclear-weapons facility for 40 years had resulted in the development of a community largely inclined to support nuclear technologies. Likewise, Yucca Mountain and Deaf Smith were located near federally-owned nuclear-weapons installations—the Nevada Test Site and the Pantex nuclear weapons assembly plant, respectively.

The controversy generated by Herrington's decision prompted DOE to commission an independent analysis of the site selection process by the National Academy of Sciences (DOE, 1986a, 1986b; Merkhofer & Keeney, 1987). This study ranked Hanford only fifth in terms of technical suitability, with Yucca Mountain ranked first and Deaf Smith third. Two other sites—Richton Dome, Mississippi, and Davis Canyon, Utah—were ranked second and fourth, respectively. It was widely expected that DOE would replace Hanford with Richton Dome, but Herrington persisted with his initial recommendations, evidently expecting more political trouble if DOE tried to characterize Richton Dome or Davis Canyon. So, on May 27, 1986, Herrington announced that he had nominated—and President Reagan had endorsed—Yucca Mountain, Deaf Smith, and Hanford. Thus, the search for a western site had been narrowed to three areas where there was a history of public support for nuclear technology.

Abandoning the Second-Round Repository

While the first-round search in the west became decidedly political in nature, the second-round search in the east unfolded much more in line with the intentions of NWPA. DOE surveyed a large number of potential geologic formations throughout the eastern United States, and then used a systematic, data-driven decision procedure to select a set of potentially suitable sites. However, there was a strong public outcry when the 12 candidate sites were announced in January 1986. And because these sites were spread over nine states (Minnesota, Wisconsin, Maine, New Hampshire, Virginia, North Carolina, and Georgia), public opposition was widespread throughout the midwest, northeast, and south.

DOE capitulated to this opposition 6 months later. On May 27, 1986, the same day the three western sites were formally selected for characterization, Energy Secretary Herrington announced that he had indefinitely suspended DOE's search for a repository in the east, arguing that a single repository would meet the country's needs for the foreseeable future. This "lack-of-need" argument was completely at odds with a statement made only a month earlier by Ben Rusche, director of the Office of Civilian Radioactive Waste Management (Carter, 1987), who had testified in Congress that a second repository would be needed.

The decision to abandon the second repository was widely viewed as a thinly-veiled attempt to enhance the reelection prospects of several senators from the eastern half of the country and to defuse political opposition to the repository from the more populous states. This assessment was corroborated by a report, prepared by a DOE consultant involved in the decision, which stated that one reason for dropping the second repository was immediate political relief (Davis, 1988).

Herrington's decision drew a rebuke from congressional supporters of NWPA (e.g., Rep. Morris Udall, D-Ariz.), who reduced the DOE budget and restricted on-site work at the three sites (Carter, 1987; Davis, 1988). This rebuke convinced Herrington to restart the second-round search, albeit with a longer timetable to completion. However, in late 1987 Congress completely restructured the failed program and passed the Nuclear Waste Policy Amendments Act (NWPAA), which officially abandoned the search for an eastern site. Section 161 imposed an indefinite moratorium on the siting of the second repository, and the search cannot be restarted until Congress specifically authorizes funding for this purpose.

Thus, one of the key guarantees of distributional equity contained in NWPA was abandoned. And NWPAA went further; not only did it relieve eastern states from being selected as repository sites in the near future, it also narrowed the first-round options in the west.

Targeting Yucca Mountain

Section 160 of NWPAA ordered a halt to all site characterization activities except at Yucca Mountain, a decision aimed at defusing the nationwide controversy that had engulfed the program. Restricting the program to a single site allowed 49 of the 50 states off the hook and thus overcame their opposition. Further, concentrating the entire effort on a single site appealed to DOE and the nuclear utilities because it made the task seem more manageable and considerably less expensive.

Of the three western sites selected for site characterization, Nevada appeared to congressional supporters of the repository program to be the most vulnerable. Nevada was by far the least populated and, with only two congressional districts, the most weakly represented in Washington, DC. Further, one of Nevada's Senators, Chic Hecht, adopted a decidedly mild approach to the repository issue, being concerned mostly with ensuring that Nevada would obtain some compensation if Yucca Mountain were selected. In contrast, Texas and Washington State were well represented in Washington, DC, and these delegations were relatively united in their opposition to having the repository in their states. The Texas presence included Rep. Jim Wright (Speaker of the House), Sen. Lloyd Bentsen (Chair of the Senate Finance Committee), Sen. Phil Gramm (a personal friend of President Reagan), and Vice-President George Bush. Tom Foley of Washington State was House Majority Leader at the time. It was not at all surprising, therefore, that when a political power play was employed, Nevada was chosen. NWPAA was attached to the fiscal year 1988 Budget Reconciliation Act and passed in the closing hours of December 23, 1987, as Congress scrambled to adjourn for the holidays.

The decision to limit site characterization studies to Yucca Mountain for political reasons violated the ethical principle of procedural equity in NWPA, which required siting decisions to be based on technical and scientific criteria. Some might argue that it is impossible to conclude that Yucca Mountain was chosen for political rather than technical reasons, because preliminary data suggested that it was the safest and least costly of the five candidate sites (Merkhofer & Keeney, 1987). However, Congress paid little attention to science in developing NWPAA, and if Congress had been interested in selecting a site on technical merits, it would presumably have delegated the decision to DOE. In fact, Mary Louise Wagner, an aide to Senator Bennett Johnston (D-Louisiana), who was chairman of the Senate's Energy and Natural Resources Committee and the architect of the NWPAA, admitted the decision was not based on technical considerations. She reportedly told Nevada Governor Miller that "the decision to target Yucca Mountain was politically motivated" (Shett-erly, 1988). Certainly, that was the widespread perception in Nevada, where NWPAA made headlines as the "Screw Nevada Bill."

The 1992 Energy Policy Act

The determination of powerful members of Congress to site the reposi-tory at Yucca Mountain was demonstrated again with the passage of the 1992 Energy Policy Act, which scrapped the standards for radiation exposure that the Environmental Protection Agency (EPA) had planned to apply in licensing the Yucca Mountain repository. In its first set of tentative standards, EPA proposed to compute these risks in terms of the cumulative release of radiation over 10,000 years. But the Energy Policy Act instead mandated an "individual-dose" standard—that is, Yucca Mountain would fail the licensing test only if there's evidence that the repository will seriously expose some individual at some particular point in the history of the facility. It is expected to be much easier for DOE to argue that a single large dose of radiation is unlikely to be released than that high cumulative releases are unlikely.

EPA officials had earlier expressed serious reservations about this change in philosophy (Galpin & Clark, 1991), whereas DOE officials had argued that EPA's cumulative-release approach was "unnecessarily conservative" (Ziemer, 1992). In acting on DOE's recommendation, Con-gress appeared to endorse the view that cumulative-release standards are too stringent. However, the EPA approach was abandoned only for the Yucca Mountain repository and was explicitly left in place for the licens-ing of the Waste Isolation Pilot Plant (WIPP), a storage facility for defense wastes near Carlsbad, New Mexico. (This disparity is particularly interest-ingly in light of the fact that the WIPP wastes, unlike those destined for Yucca Mountain, are transuranic and considerably less lethal than HLNW.) This change suggested to many observers, especially within Nevada, that the shift in approach for Yucca Mountain was designed primarily to increase the chances of licensing that site.

Senator Bob Graham (D-Fla.), whose own state would benefit by having the repository in Nevada, expressed concern that Congress was changing the siting rules because DOE anticipated that the proposed facility might have problems meeting EPA's original radiation exposure standards. He strongly argued that it was scientifically unsound to devise a set of licensing standards after a site had been selected, and then apply those standards only to that site. And he castigated Congress for trying to maintain the licensability of a questionable site, "This approach will destroy any remaining public confidence in the integrity, fairness, trust, and worthiness of the federal government in carrying out its responsibilities with respect to nuclear waste. The message from this legislation is simple: The federal government will do anything; it will say anything; it will spend an unlimited amount of ratepayer dollars; and it will make up the rules as it goes along, including the standards for the protection of the public health and safety, in order to find as quickly as possible a place to dispose of or to store HLNW. Any state, scientific viewpoint, fact, or law that becomes an obstacle to this objective will be legislated out of the process" (Graham, 1992, p. S17577).

Lack of Funding for State Studies

Another series of events reinforces the conclusion that the ethical principles for site selection contained within NWPA have been largely abandoned. As mentioned before, NWPA provided for candidate states to receive federal funding to conduct independent studies, a provision that could mitigate the concerns of the host state about a repository. However, Nevada's ability to conduct scientific studies of the suitability of Yucca Mountain has been compromised by Congress' deliberate lack of funding, at least since 1988.

Following NWPAA, Congress sought ways to prevent Nevada from blocking the repository. A primary target was the state oversight program; state scientists were uncovering evidence that Yucca Mountain might not be suitable, and the state had established a high-profile agency—the Nuclear Waste Project Office (NWPO), which funded the studies on which this book is based—to disseminate its concerns both within the state and to neighboring states with transportation corridors. In an attempt to stem public opposition, Congress has repeatedly cut federal appropriations for Nevada's oversight of the repository program and, at the same time, imposed restrictions on the type of activity that can be supported with oversight funds.

In conclusion, Congress and DOE have conducted themselves in a manner that suggests to Nevada residents that building a repository at Yucca Mountain is the paramount objective, regardless of site suitability. Any intent to identify the technically-best site was abandoned by 1986; indeed, almost all the provisions of NWPA designed to ensure a fair siting process have been either rescinded or severely compromised.

These violations of procedural equity energized Nevada's opposition to the repository. In fact, an editorial in the Las Vegas *Review-Journal* suggests this violation is the primary objection of state residents:

A statewide Review-Journal poll . . . [shows that] fully 77% of Nevadans flatly oppose the federal plan to bury the nation's nuclear garbage at Yucca

Mountain . . . . In these figures, surely, there is a strong element of what some call *nuclearphobia,* an irrational and unjustified fear of anything and everything associated with atomic power. But the poll results also reflect a feeling of resentment on the part of Nevadans. The people of this state realize that Yucca Mountain was not selected on the basis of hard science, but rather on the basis of hardball politics. The key issue, as many Nevadans, including ourselves, see it, is fairness. (October 30, 1990)

Given that opposition by Nevada's residents and elected officials has been a major obstruction to site characterization of Yucca Mountain, we might conclude that the failure of DOE and Congress to adhere to the statutory standards established in 1982 is one cause for the current failings that plague the repository program.

It is highly unlikely that the current HLNW policy will bring a repository on line by 2010 (GAO, 1993). One reason is the abandonment of the original guiding legislation. On the other hand, even if Congress and DOE had adhered to NWPA, it is not at all clear that a politically and technically suitable site would now be identified. Indeed, the second-round search for sites in the eastern part of the country suggests that NWPA might not have succeeded even if it had been faithfully played out. Although DOE engaged in a systematic and objective investigation of potential sites for the second-round repository, the selected states responded with a level of outrage that brought down the entire program.

Despite repeated congressional attempts to clear the way for DOE, the agency faced a set of hurdles that make successfully siting a repository difficult under any circumstances. Three key obstacles stand in the way: the tremendous scientific uncertainty inherent in the geologic disposal of HLNW; the deep concerns expressed by the general public regarding HLNW disposal; and ill-equipped management structures and short-sighted decision-making on the part of institutions responsible for dealing with HLNW.

Unraveling of the High-Level Nuclear Waste Program

The 1982 NWPA, which was intended to lay a solid policy foundation for the national HLNW disposal program, arose out of a need to break a series of political deadlocks that blocked development of a national HLNW management program. NWPA placed exacting demands on the federal government and especially on DOE, including the requirement to establish an administrative and managerial structure capable of implementing NWPA's equity provisions. Unfortunately, DOE was unable or unwilling to comply; indeed, many of its actions were clearly calculated to circumvent the intent of NWPA. With NWPAA, Congress and DOE all but abandoned the opportunity and obligation to create the governmental relationships necessary to implement a successful waste management policy and to resolve conflicts arising from political, administrative, and management issues. In short, they utterly failed to lay the institutional groundwork for satisfactory state-federal relations that could facilitate repository siting and development.

To understand the nature and magnitude of this failure, it is useful to examine the issue of political control of the process for selecting the repository site—in particular, the respective roles of the federal government and its agencies on the one hand and that of states and Indian tribes on the other.

As we have seen, the intent of NWPA was to ensure fairness in the site selection process. It divided responsibilities between the federal and state players and—along with the congressional discussions and hearings that preceded its passage—created expectations about federal policy, states' rights and powers, and the relationship between the federal government and the states. These expectations are critical in understanding the impact that subsequent deviations from the policy had on the states' perceptions of the legitimacy of the federal government's actions in general, and DOE's in particular.

Let us examine first the responsibilities and rights accorded to the states. Before 1982, a long and troubled history of relations between the states and the federal government over nuclear waste disposal issues had already shaped the perceptions and expectations of everyone involved. As early as 1976, DOE's predecessor, the Energy Research and Development Administration, had begun to search for a repository site in 36 states, disregarding both public and state participation (Carter, 1987; Kraft, 1992). The exclusion of local participation provoked two-thirds of the states to enact regulations covering various elements of HLNW management, and several states went so far as to ban any site exploration within their borders (Kraft, 1992). And the issue of nuclear waste management and disposal was only one of several that caused the relationship between DOE and the states to deteriorate to the point of open hostility. Another major area was management of the nuclear weapons complex (Dantico & Mushkatel, 1991; Shulman, 1992).

Despite this adversarial situation, however, there were some developments that increased expectations that state and public participation would be encouraged and that the search for disposal sites would be conducted in a fair and nonpolitical manner. Notably, President Carter's Interagency Review Group engaged in what Kraft (1992) characterized as a significant effort to "promote policy consensus" (p. 154). President Carter's efforts held the promise of consultation and concurrence with the states and of an on-going role for states throughout the repository development process. President Carter also created an Executive Planning Council to examine site selection procedures, "in consultation with the states" (Kraft, 1992, p. 154; Jacob, 1990, p. 80).

Just before NWPA was passed, however, the Office of Technological Assessment (OTA) issued a report that, while not critical of whether disposal technology could be developed, did express severe reservations about the ability of DOE to administer and manage a sustained program. Moreover, OTA believed the proposed legislation did not adequately define the institutional arrangements crucial to the success of the program, and suggested that the lack of trust among the concerned parties "threatened to lock the waste disposal effort in a state of virtual and continual paralysis" (OTA, 1982, p. 95).

This criticism was repeated in many subsequent analyses of the program, including those by the DOE Advisory Panel on Alternative Means of Financing and Managing Radioactive Waste Facilities (DOE, 1984), by GAO (1993, 1994), by the National Research Council/National Academy of Sciences (1990), and even by the U.S. Secretary of Energy's own Advisory Board Task Force (1993). (The Advisory Board was in itself an extraordinary creation, charged with the monumental task of recommending ways for DOE to regain public trust and confidence.)

Jacob (1990) believed NWPA represented a reassertion of the nuclear establishment's preeminence at the expense of competing proposals that would have succeeded in reconstructing the nation's nuclear waste policy. Hence, NWPA ultimately failed because its inherent design, which reflected the existing political distribution of power, did not bring about the kind of institutional changes that would ensure acceptance of the policy by the states and the public. This is not to underestimate the difficulty of the task. As Kraft (1992) noted, "How can the nation create a process for repository siting that includes extensive state and public participation (especially with a state veto) and yet adhere to a scientifically defensible site evaluation and selection procedure that will result in construction of a repository?" (p. 155).

One of the major flaws in NWPA concerned the roles assigned to the states and the public in the site selection process. While NWPA recognized the importance of working with both states and Indian tribes and afforded them legal standing, it nevertheless reserved to Congress and its committees all of the real power to oversee administration and implementation of NWPA. NWPA allowed states to conduct their own socioeconomic studies, but it did not require that the findings play a determining role in site evaluation, nor could potential social or economic impacts be used to disqualify a site. States could monitor and evaluate DOE's scientific work but they were assigned no decision-making role. States could litigate and otherwise oppose site selection and development, and they even had a limited veto power. Ultimately, however, Congress could still force a state to accept a repository because any state veto could be overridden if a majority in both houses of Congress supported the selected site within 90 days of the state's action. This meant that a state veto had very little chance of succeeding, because it would be extremely difficult for a single state to gain majority support in one of the houses of Congress. Congressional delegations would align themselves with DOE against an individual state, as long as they were convinced that building a repository was inevitable and they wished to avoid the political liability of having it sited in their state. They might also be motivated by other factors, such as a desire to "solve" the problem of HLNW disposal, to hasten the removal of wastes stored at reactor sites in their states, to respond to the lobbying of the nuclear power industry, and to obtain some results for the tremendous costs of the site characterization program.

Recognizing the extent of congressional authority over implementation of NWPA, many states quickly established their own oversight agencies to keep an eye on DOE's activities in their state.

NWPA required DOE to consult with potential host states, but, as Jacob (1990) pointed out, NWPA did not require DOE to obtain the states' concurrence with its actions. And, all too frequently, DOE interpreted consultation to mean simply giving notice of an impending action, or worse, to engage in a promotional public relations effort. The lack of any requirement in NWPA for DOE to engage the state in meaningful negotiation or arbitration suggests that Congress would likely make the final decision. Therefore, as long as DOE could maintain congressional support for its implementation of the repository program, it could largely ignore any objections from potential host states. With Congress as the final arbitrator and the states in possession of no real power, NWPA failed to provide any meaningful "court of last resort" outside the judiciary. It was clear that decisions about the repository program would be based more on congressional politics than on technical considerations—a situation that did nothing to make DOE's job easier.

NWPA also failed to give states power to deal with another important aspect of the waste disposal problem: transportation. Studies have shown that the public perceives high risks to be associated with transporting HLNW (Mushkatel & Pijawka, 1992; Pijawka & Mushkatel, 1992). Yet NWPA made no provision permitting states to challenge Department of Transportation regulations governing waste shipments (Jacob, 1990).

In summary, then, while NWPA set admirable goals for creating a relationship between state and federal authorities that would expedite repository siting, NWPA failed to lay an adequate institutional foundation for realizing those goals. It gave states and Indian tribes no final authority on any issue of substance. And, unfortunately, NWPA assigned the responsibility of implementation to DOE, an agency that few people or states trusted, to develop a fair siting process or to resist politically-expedient solutions when it found the going tough (OTA, 1982).

Nuclear Waste Policy Act Amendments:
An Intergovernmental Crisis

Two years after President Reagan signed NWPA, it had already started to unravel. As we have seen, federal political objectives came to dominate the search for the repository site, largely pre-empting the role of the states. DOE's disregard for the equity provisions of NWPA made it inevitable that the candidate state finally chosen would fight what was coming to be regarded as a biased and fatally flawed selection process and a blatant violation of its rights under NWPA. The passage of NWPAA made it clear the federal government would, if it could, impose the selection of Yucca Mountain on Nevada, regardless of the resistance it encountered. Jacob (1990, pp. 162–163), concluded, "In sum, little evidence suggests that the Yucca Mountain, Nevada, site was the outcome of a systematic site selection process. Rather, it should be seen as the product of historical precedent and political-economic expediency."

It was widely believed in Washington, DC, that while Nevada had been the victim of congressional politics and an inequitable siting process, the site selection was nevertheless a done deal. Yet, such a perspective is myopically federal; in fact, Nevada's resistance has been poorly documented (or understood) by those who view the program from the federal perspective, and consequently its political importance has been underestimated.

State governments do have a wide variety of powers that can be used to delay and ultimately impede implementation of federal policies. It was only to be expected, therefore, that Nevada would use whatever legal means it could muster to fight the federal government's imposition of a repository site on the state.

Nevada's Response

Two years before NWPAA was passed, Nevada had already established its own Nuclear Waste Commission (NNWC) and NWPO. NWPO was to provide the studies that would allow the NNWC to independently assess the suitability of the Yucca Mountain site and allow the governor to make recommendations to the president and Congress on the matter. Early interviews with state administrators in several agencies indicated there was a low level of awareness about the repository and few planning activities were being devoted to the siting process (Mushkatel, 1987). Prior to the passage of NWPAA, there was no evidence of overwhelming opposition to the repository, according to interviews conducted with numerous state and local governmental agency directors, city managers, and local elected officials in Clark County, including at least two city councilors and two commissioners from each municipality and every mayor in the metropolitan area. However, these respondents noted that a marked shift in opinion occurred after NWPAA was passed (Herzik & Mushkatel, 1988, 1991/92).

As Jacob notes (1990), one goal of NWPAA was to undermine the state's opposition to the siting. It not only included financial incentives to accept the site, but also established a grant program for affected counties to conduct studies, separate from the state's program. At the same time as these local programs were being set up, the funds allocated to the state agency, NWPO, were cut drastically—an extension of previous federal efforts to fragment opposition to DOE's programs. This cut was the first of many such funding cuts, all of which DOE attributed to Congress.

The response of local officials to NWPAA was telling. The mayor of Las Vegas, one of several who visited Washington, DC, prior to the passage of NWPAA, to obtain first-hand information, reported that Representative Udall had told him that the repository was a done deal and to put his wish list together. Yet, even though the great majority of these state government leaders believed the repository to be inevitable, not one believed DOE could be trusted to carry out its promises (interview, May 1988). Indeed, by April 1988 Clark County had a resolution seeking DOE funding as an affected county so it could conduct its own studies.

Meanwhile, the Nevada legislature was also expressing its displeasure with the conduct of the repository program. During the 1989 session (the first meeting of legislators following passage of NWPAA), the legislature unanimously passed two resolutions strongly opposing the repository and formally denying the state's approval for ceding jurisdiction over the land required for the facility. Taken together, these resolutions were as strong a statement of opposition the legislature could make. Transmission of these resolutions to Congress provided the foundation for Nevada's claim that it had, in fact, exercised its veto of the repository under NWPA. Then, toward the end of the session, the legislature passed Assembly Bill 222, which made it illegal to dispose of HLNW in Nevada. Only six state senators voted against it.

An opinion by the Nevada attorney general supporting the legality of these laws (McKay & Swainston, 1989) led Governor Miller to write to the U.S. Secretary of Energy and ordered him to cease all site-characterization activities. The secretary responded on November 29, 1989, with a report to Congress indicating that site characterization should proceed (DOE, 1989). Nevada's governor and attorney general then pursued two distinct strategies to enforce the state ban. First, Governor Miller ordered the state engineer and the Division of Environmental Protection not to act on four DOE applications for environmental permits; then the attorney general filed suit against the U.S. Secretary of Energy, arguing that the secretary did not have authority to consider the Yucca Mountain site. Nevada presented three arguments in favor of disqualifying Yucca Mountain: (1) the three anti-repository bills constituted an official "notice of disapproval" (a state veto authorized under NWPA) and Congress failed to respond within the required 90 days; (2) a state has the authority to regulate HLNW within its borders if the regulation is based on economic purposes; (3) NWPAA authorizing site characterization of Yucca Mountain was unconstitutional on a number of grounds. However, the United States Circuit Court of Appeals (designated by NWPA as the court of original jurisdiction for the repository program) ruled against Nevada, finding that the federal government had the right to proceed with its study of the Yucca Mountain site (*Nevada v. Watkins*, 1990). This decision was left standing when the Supreme Court elected not to hear the case.

Although the court decision required Nevada to act on DOE's permit requests, the state found another way to delay the process. The attorney general contested DOE's application to the state engineer for a water permit, which was necessary to conduct various site characterization activities. Although the permit was eventually granted—though only for a limited time—it was not issued until nearly 4 years after DOE originally made its request.

It is important to note, however, that DOE was successful in some of its undertakings in Nevada. The implementation of NWPAA successfully played on traditional intergovernmental tensions between northern and southern Nevada. There has been some fragmentation in the state's position, and the pro-repository lobbyists have probably made inroads with some state legislators. Nevada citizens, however, remain adamantly opposed to the siting (Mushkatel & Pijawka, 1992; Flynn, Slovic, & Mertz, 1993).

It is clear that the relationship between DOE and the state severely limits any meaningful negotiation. Since the passage of NWPAA, DOE has attempted to force Nevada's compliance either by using or threatening sanctions such as budget cuts. And when that has not worked, DOE has resorted to using rewards in attempts to split and fragment opposition within the state. For example, research funding for state universities is an odd mixture of direct DOE contracts and indirect congressional grants, the latter coming via funding to NWPO earmarked for use by the universities. On the direct DOE funds, NWPO and the state universities have disagreed vehemently over whether the acceptance by the universities of direct funding from DOE through the vehicle of cooperative agreements could be construed as an indication of the state's willingness to cooperate with the federal Yucca Mountain project, thereby compromising the state's legal and moral position.

Despite its efforts, however, DOE has made extremely little progress in obtaining Nevada's cooperation or its acceptance of the repository. What DOE has not attempted is an effort to develop new institutional roles for the state and the public in the repository siting process. Despite extraordinary delays in this program, DOE personnel seem completely unable to accept—as Morris Udall did in 1987—any suggestion of a moratorium and reconsideration of the existing institutional relationships (conversations and discussions with DOE personnel, February 17, 1993).

Although Nevada's opposition has played an important role in impeding DOE's progress, it is important to note that its political and legal challenges have not been solely responsible for delaying the repository program. In many ways, DOE has been its own worst enemy; it has reorganized and rescheduled the program so often that deadlines can no longer be taken seriously. The incredible shifts in the projections for completion of the project have contributed to the declining confidence in DOE's competence to manage the program (U.S. Secretary of Energy, 1993).

In response to DOE difficulties with the Nevada public and stakeholders, the American Nuclear Energy Council (ANEC) decided to provide its support to solve these problems. As an association that represents the interests of the nuclear power utilities, ANEC wanted to do all it could to promote acceptance of the repository. Ambitious plans were made for a public relations campaign to influence Nevada public opinion and a high-pressure lobbying campaign to influence state legislators. Part of this plan was to train DOE scientists and program officials as *truth squads* to counter the claims and statements of repository opponents.

The public advertising campaign was not successful, and in its original format was discontinued a few months after it began. Survey data showed that more people opposed the repository than supported it in response to the ads. At the same time, a confidential document outlining the strategies of the ANEC effort was leaked to the press, and the entire plan became a controversial issue in Nevada (Flynn, 1992; Flynn, Slovic, & Mertz, 1993).

The attempt to lobby the state legislature during their 1993 session was equally unsuccessful. The plan taken into the legislature was watered down to a rather mild resolution in support of examining possible benefits the state might obtain from the federal government in return for studying Yucca Mountain. The State Senate was thought to be the most likely to support the resolution, but when it was brought to vote late in the session the state senators voted it down by a two to one margin (German, 1993).

The attempt of ANEC to team up with DOE and influence the political process in Nevada—through an advertising campaign to change public opinion and through a massive lobbying effort—has left only a record of yet another failure in obtaining Nevada's support. The nuclear power industry has demonstrated that its understanding of the situation and its idea of a solution through influencing intergovernmental relations between the state and the federal government is as uninformed and mistaken as that exhibited by DOE.

4

Complexity and Uncertainty

Certainty cannot be the prior goal for human action in the management of complex technologies. However, a wise public policy must clearly face the implications of uncertainty and provide reasonable assurances of safety to human health and the environment. Such policies should not deny the potential consequences of human lack of foresight nor become paralyzed into inaction. If there are risks to be taken under conditions of uncertainty, these risks must be acceptable to those citizens, communities, and states who will bear the consequences.

The Problem of Complexity

Building a high-level nuclear waste (HLNW) repository is, in itself, one of the most complex technological ventures the United States has ever undertaken, but it will be only one component of a vast, nationwide, waste management system. The nuclear power industry now holds spent fuel rods at 72 sites from about 110 reactors. The Department of Energy (DOE) operates a 14-site defense weapons complex involving 13 states. Several of these sites contain large amounts of HLNW, in some cases stored in unstable conditions needing immediate, interim, and long range attention (OTA, 1991).

The elements of the repository program have started to evolve, and it now comprises a multilayered and geographically dispersed enterprise that involves many players: DOE, Congress, the nuclear power industry, private contractors, national laboratories, the United States Geological Survey, a host of subcontractors, and a multitude of advisory and review committees. Some 2000 people work on the program, 90% of whom are not DOE employees. Management turnover has been high; the program has had seven directors in the past decade, five of which never moved beyond the status of acting director (NWTRB, 1993). Simply keeping the management of this complex operation up and running consumes more than half the program's funds (Edison Electric Institute, 1992; GAO, 1993).

The future holds many crucial questions: What type of transportation system should be used? What role should storage at reactor sites play? Should a monitored retrievable storage (MRS) facility be part of the system? Should engineered barriers or geology be the primary means of isolating the wastes? What types of casks should be employed? Should wastes be sequestered in the repository as rapidly as feasible or purposely allowed to remain accessible in anticipation of future developments in scientific knowledge and technology? What is the future of nuclear power, and what volume of wastes will it create?

Many considerations impinge on the solutions to these questions: How will the solutions impact the nuclear industry? Will new storage and disposal alternatives arise over the next few decades? Will various publics become more—or less—accepting of waste disposal facilities? What positions will be taken by future Congresses and political administrations? Will a cure be found for radioactive-induced cancers, thus, perhaps, reducing concerns about potential health impacts of HLNW?

All of these questions must be addressed, and their components integrated into the development of a smoothly functioning national waste management program. However, to date relatively little attention has been paid to the entire daunting scope of this challenge. Instead, most of the effort has focused on a few relatively narrow issues associated with building the first HLNW repository: coming up with an acceptable design, finding a suitable geological disposal site, creating a strategy, and formulating environmental standards and regulations to ensure long-term isolation of the wastes from the biosphere and nearby populations. DOE has been expected to shoulder unfamiliar tasks: managing the far-flung repository development project, licensing a first-of-a-kind HLNW disposal facility, and steering the overall enterprise with an exemplary level of performance that rises above their past failures in handling defense wastes.

Given the problems that have so far confronted DOE in its effort to site a repository, it is not surprising that the larger issues have not been addressed. But the controversies, delays, public mistrust, and political conflict that have characterized the initial phase of the exercise do not inspire optimism. The road ahead will not be easy. Complexity and uncertainty seem destined to remain on the agenda; the repository program should be viewed more as an experiment than as an established technological undertaking.

The Problem of Uncertainty

Time—the future—is the most formidable challenge to the success of the disposal program. HLNW must be isolated for 10,000 years, twice the span of recorded human history. DOE must predict and convince all its reviewers that basic safety will be achieved over this period. At the same time, predictions of social and economic change are notoriously unreliable even over periods as short as 5 years—as the fate of most 5-year economic plans attests. It's to be expected, therefore, that over periods exceeding 50 to 100 years—and certainly over 10,000 years—unforeseen events and changes in the natural environment and human culture will occur, however carefully we try to plan for them. Such changes may directly affect the Yucca Mountain disposal site, fundamentally altering key variables such as population growth, the effectiveness of engineered barriers, and groundwater flow.

But accepting that changes will doubtless occur does not mean we can tell exactly when they'll occur, or what effects they will have. The intrinsically unpredictable nature of surprises makes them impossible to anticipate using current analytical methods (such as computer-based projections or simulations), even in calculating uncertainties. This inability greatly limits the usefulness of probabilistic risk analysis in dealing with a long time frame characterized by a high percent of surprises—precisely the situation that applies to a underground, long-term, HLNW repository.

As noted previously, geological isolation of HLNW can be considered a form of reverse mining, in that something is put into the mine rather than extracted. Because mining is fundamentally an exploratory activity, miners often encounter surprises, and techniques have been developed for safely extracting ores even when conditions underground are very different than expected. In contrast, the United States waste disposal program has taken the approach of developing a complex regulatory system involving highly specific standards, criteria, and regulations. These rules demand the ability to predict in detail the behavior of rock for 10,000 years into the future, as well as the future behavior of the stored HLNW, the waste canisters, the repository's sealing system, and so forth. Such predictions inevitably rest on many assumptions, are uncertain to varying degrees, and "will remain uncertain no matter how much additional information is gathered" (NRC/NAS, 1990, p. 2).

So while risk and performance assessments may be helpful in identifying deficiencies in the repository's isolation system, in improving the nature of engineered barriers, and in designing monitoring programs, these assessments have only a very limited ability to predict the probability and magnitude of potential dangers associated with the HLNW repository over 10,000 years. The assessments can provide partial and useful information about what events might occur and when, and what their ramifications and consequences might be, but the inherent level of uncertainty in the undertaking will continue to defeat the ability of such assessments and scientific studies to "prove" that a repository will be "safe" over such a long time. Given this uncertainty, citizens and their elected leaders will ultimately weigh the best information science can provide and make a judgment, based on social values, about whether to proceed (NRC/NAS, 1990).

But an unknown future is only one source of uncertainty. Several others have been enumerated by Sven Oye Hansson (1992), a member of the Board of Directors of the Swedish National Board for Spent Nuclear Fuel (Statens Kärnbränslenamnd; SKN). These sources include uncertainties of consequences, values, demarcation, reliance, and coordination.

Uncertainties of Consequences

It is possible to anticipate many possible outcomes and consequences associated with different waste disposal methods or alternative types of engineered barriers. What is not well known, however, is the probability that any of these outcomes and consequences will actually happen. And today's experts cannot reasonably claim to have thought of every possible consequence that might occur. Thus, estimates of the overall risk associated with HLNW disposal must necessarily be based on an incomplete list of possible consequences. These estimates may be viewed as decision making in the face of unknown possibilities.

Uncertainties of Values

The choices we make, individually and as a society, are guided by social values. Likewise, outcomes or consequences of those choices are measured in terms of values, such as human well-being, ecological diversity, or money. But because individuals, groups, and organizations often weigh these values in different ways, many social decisions and choices become difficult and contentious. This difficulty is particularly true of choices about HLNW disposal because extraordinary uncertainties are associated with the values to be applied. It is not just that different participants in the decision-making process have diverse views as to which values should take precedence. There is also the fact that the current generation is ignorant about the values of future persons. Will they, for example, require a level of protection against radiation hazards that is stricter, the same, or more lenient than that currently in existence?

If history is any measure, two things are certain: the values of future generations will be different from those today, and we have little ability to foresee what they will be. When the short span of the past 50 years is considered, it is apparent that public values in the United States regarding technology, the environment, health, and the trustworthiness of institutions have undergone remarkable changes. People living hundreds or thousands of years from now are likely to attach great importance to social goals and values that this generation may not have even recognized. Considering the current pace of change in public values, our limited abilities to formulate social policies to which succeeding generations will not object must be recognized (Hansson, 1987). This limited ability is a problem that besets much environmental (and other) decision making, but it is particularly acute for HLNW disposal. And current plans must take into account that measures specifically taken to build a better future carry the substantial risk that these measures may have the opposite effect.

Uncertainties of Demarcation

Different groups and organizations view the focus, scope, and time frame of the HLNW disposal problem in different ways, and it is not clear which view will prevail. Disposal experts concerned primarily with the wastes generated by the current generation of reactors have the shortest horizon. Nuclear industry representatives, in contrast, often see an energy production horizon extending into the future and involving a second generation of reactors. Environmentalists and critics often view the waste problem in a broader energy or social system context and pose the question: What does HLNW disposal suggest about the preferred way to organize society and to produce and consume energy? Each context can provide a basis for rational decision-making, but uncertainties abound as to which provides the appropriate scope of analysis or set of decision norms.

Uncertainties of Reliance

The technological complexity of HLNW disposal forces reliance on scientific research and technical assessment to develop knowledge and expertise. But is it rational to rely so heavily on technical experts, particularly those involved in the development task itself? Increasing reliance on scientific and technical experts has the effect of further removing from the decision-making process the public and stakeholders who ultimately bear the risks. This distancing, in turn, increases the uncertainty that such reliance on experts will ultimately be accepted. Evidence of this unacceptance can be seen in the increasing militancy of those potentially affected by technological programs, who demand to have their concerns heard and considered by authorities, and who simply reject attempts to override these concerns by recourse to expert opinion.

Those who have privileged access to such expertise can be insensitive to this type of uncertainty; indeed, they are often baffled by what they perceive as scientifically-unsophisticated emotionalism and irrationality on the part of public.

Uncertainties of Coordination

Because deciding what to do about HLNW is ultimately a social decision to be decided by citizens as well as government agencies and experts, a successful program requires development of a decision-making process and an organizational structure that can accommodate and coordinate the views and actions of groups with different preferences and ethical norms. Some, for example, want the problem solved now so that it will not be exported to future generations. Others prefer the lowest-cost solution, whatever its time frame. Still others wish to postpone the final decision to avoid possible political repercussions or to allow scientific knowledge to increase and technical uncertainties to be resolved. How to mesh these different objectives and norms, each with a competing and sometimes irreconcilable social agenda, into a coherent social decision is a key uncertainty. A socially acceptable decision-making process does not exist for considering the range of possible alternatives and making suitable tradeoffs, compromises, concessions, and outcome decisions. Some results of this lack of process is that stakeholders become adversaries and the federal program is constantly threatened with collapse.

Uncertainty at Yucca Mountain

In each area the unknowns are formidable. Consider, for example, the assumption that the steady application of scientific research and risk assessment studies can reduce the technical uncertainties associated with the waste disposal program. This is the approach on which the assessment and planning program is predicated and on which the safety of the disposal program will be sold. But the experience with one central technical issue, modelling groundwater flow through fractured rock, illustrates how illusory is this quest for shrinking uncertainties.

Scientific Uncertainties at Yucca Mountain

According to the National Research Council/National Academy of Sciences (1990), an ambitious two-decade effort on this problem has led to an understanding that the phenomena involved are more complicated than earlier believed. Rather than decreasing uncertainty, the research has "increased the number of ways in which we know that we are uncertain" (p. 4). The Swedish HLNW research program reached a similar conclusion: Despite the expenditure of substantial money and manpower, "the experience seems to be that...we will never get more than fragmented knowledge" (Rydell, 1992, p. 20).

These results are in line with a review of nuclear risk assessment in five countries that found that risk assessment rarely reduced debates over the risks involved but, in fact, often identified new risk issues and enlarged the scope of the debate (Kasperson & Kasperson, 1987).

A second technical issue concerns future seismic activity at Yucca Mountain. It is not just a matter of trying to predict time, place, and magnitude, but also to estimate the confidence that can be attached to these predictions. In the case of Yucca Mountain, four specific impediments to prediction exist:

1. The amount of seismicity prior to 1960 is essentially unknown. The study area has always been sparsely populated and thus not of great interest for predicting earthquakes. Also, before 1960 the monitoring equipment used at the site would not be able to pick up low-level but frequent activity.

2. The effects of nuclear testing on faults within and near the site are uncertain. Shear fractures have been induced by underground nuclear explosions, and it is possible that tests have activated so-called inactive faults. But without records this activity is uncertain.

3. The long-term future entails unavoidable uncertainty because areas with a seemingly low level of seismicity can have unexpected events. In fact, strong earthquakes occur in new places as often, perhaps even more often, than in conventional, well-known places. On June 29, 1992, a magnitude 5.6 earthquake, believed to be the largest in this region, occurred about 12 miles southeast of Yucca Mountain, causing $1 million in damage to DOE's above ground facilities. Moreover, the timing of the quake, right on the heels of a much larger 7.4 magnitude event the previous day at Landers, California, suggests possible (but ill understood) links between the two quakes. Thus, additional uncertainties are introduced by the possible occurrence of multiple earthquakes and their potential to affect the integrity of waste containers and above ground facilities (especially *hot cells*, special rooms in which wastes are remotely handled and repackaged) prior to repository closure. Some of these issues have received only limited analysis (NWTRB, 1992).

4. The methods and techniques of geological science were developed to explain and understand events that have already occurred. They have been pressed into service to predict. This attempt is, according to Massachusetts Institute of Technology geologist K.V. Hodges (who reviewed work done for DOE as part of the *Early Site Suitability Evaluation*), a hit or miss proposition. He added that tectonic predictions, when stripped of statistical sound and fury, are not much better than educated guesses, and that asking for predictions of seismic events for 10,000 years in the future is asking for the impossible (Shrader-Frechette, 1993; Younker et al., 1992).

Social/Cultural Uncertainties at Yucca Mountain

An entirely different set of uncertainties is associated with potential future human intrusion into the Yucca Mountain site. It is not difficult to envisage credible events over succeeding centuries that could result in unanticipated human intrusion.

Military activities such as bombing and tests. Historically, there has been an average of four major wars per century worldwide, and the United States has experienced about two per century. Although the frequency of wars may decline because of the deterrence effects of modern weapons or international agreements, it would be unduly optimistic to assume that all military activities will cease over the next 10,000 years.

Underground settlement. Dramatic changes in housing patterns and urban settlement have occurred over the past few centuries, so it is not unreasonable to assume that similar radical changes will occur in the future. Population increases and housing demand may make underground settlements more feasible (or desirable) than alternatives to surface dwellings, such as ocean habitats or space colonies.

Underground traffic. If society continues to permit individual means of transportation, such as the automobile, crowding on surface routes may force extensive development of underground alternatives, such as those that have already become common in some large cities. Urban development in southern Nevada may also one day necessitate development of an underground public transportation system. Although at present it may appear that Yucca Mountain is unlikely to attract an urban population, it is entirely possible that some of the region's traditional constraints to growth, such as climate, aridity, and topography, may be overcome over hundreds and thousands of years. The area might offer new economic advantages, such as clean air or access to space travel.

Intrusive large-scale landscaping. If urban development increases, especially if accompanied by underground transportation systems, new forms of regional landscaping may evolve. Deep storage of HLNW may not seem deep at all if the intervention of future generations matches that which has occurred to date. Other possibilities include resource development near the repository and perhaps even intentional mining of wastes, should future generations develop means to use them as resources.

It is clear that a high level of uncertainty will pervade our efforts to dispose of HLNW. Many sources of uncertainty, both technical and social exist. For example, geology is not a predictive science, therefore uncertainties concerning the performance of the geologic barrier isolating wastes from the environment will prevail. Furthermore, unanticipated interactions are likely to occur among the wastes, the engineered barriers (such as storage casks), and the surrounding environment and groundwater. A new generation of containment casks must be designed, built, and maintained, and there is as yet no way of knowing how well they will perform. Given the long time periods involved, uncertainties about the future social, demographic, and technological environment affecting the repository site will continue. Because the repository is a first-of-a-kind facility, inherently irreducible uncertainties are associated with this unique development. Finally, although it is often assumed that further scientific research and risk assessment will reduce the uncertainties, it's actually more likely that these uncertainties will continue to grow.

And the uncertainties do not end there. They also play a role in comparisons between the current policy—long-term geologic disposal—and alternative strategies, such as an MRS facility or storage at nuclear reactor sites. Comparative analyses of these strategies have not been a regular part of the repository assessment and management program, but one analysis was recently done by Keeney and von Winterfeldt (1994). Their study included such uncertain events as a cure for cancer, possible alternative uses for HLNW and technological innovations in waste management. Keeney and von Winterfeldt concluded that the direct economic costs of geologic storage would exceed an MRS facility or on-site storage by more than $10 billion. What matters about their report is not so much whether the analysis is correct but the fact that it underscores the urgent need for a means of assessing a continually changing set of disposal options, all based on different and highly uncertain assumptions.

A final and extremely important source of uncertainty stems from a now widely-recognized loss of public trust in the institutions managing nuclear wastes. The United States Office of Technology Assessment (1982) judged that the greatest single obstacle to a successful waste disposal program was "the severe erosion of public confidence in the federal government that past problems have created" (p. 10). The Nuclear Waste Technical Review Board (1992) and the National Research Council/National Academy of Sciences (1990) have forcefully affirmed that judgment. A special panel established by DOE to examine the trust issue has detailed the magnitude and depth of the problem and made a number of recommendations for action to combat it (U.S. Secretary of Energy, 1993). But the panel forthrightly acknowledged that DOE currently lacks the institutional capacity to strengthen public confidence. More fundamentally, it is not known the extent to which social trust, once lost, can be regained. The remedial measures are not well understood either. This area of uncertainty will be further discussed in chapter 5.

A New Management Approach

Managers in institutions confronted with such high levels of uncertainty react in various ways, most of them counterproductive: firing the bearers of bad news, obfuscation, repression, reassurance, and fire-fighting. If no way to reduce or eliminate uncertainties exists, managers often respond by repressing their awareness of uncertainties, or by treating the situation as more certain than it is.

Even more alarming are attempts to treat ignorance and uncertainty as positive conditions in making risk assessments. This happened in the DOE-sponsored work, *Early Site Suitability Evaluation*, where the DOE study team structured the problem in this way: "If...current information does not indicate that the site is unsuitable, then the consensus position was that at least a lower-level suitability could be supported" (Younker et al., 1992, p. E-11). Shrader-Frechette (1993), in her detailed review of this document, pointed out that this appeal to ignorance does not allow any rational decision on suitability. The subsequent statements by a manager of the Yucca Mountain program that this document demonstrated the repository was on the right track was simply mistaken, and unfortunately the statement came from a trained engineer assigned to oversee the complex scientific work needed for site characterization. This report and the DOE endorsement was widely reported.

Another pervasive but undesirable response is to couch ignorance or uncertainty in terms of risk—that is, to act as if there's sufficient knowledge to assign probabilities to events and outcomes. This assignment of probabilities to situations in which they are not justified is a form of self-delusion and a means of deluding others. In reference to DOE's program, the National Research Council/National Academy of Sciences (1990) has described it as an inappropriate application of science.

As a result, information or assessments that challenge current wisdom are often ignored, and the assessors end up trying to defend models and analysis techniques that are being expected to accomplish things they really are not equipped to do. But is it possible to establish a management system that can deal with uncertainty rather than trying to avoid and suppress it? In addition, can the public be educated to have more realistic expectations about how much certainty can be achieved? The most constructive approach is to acknowledge that uncertainty exists and to share that reality explicitly with government officials, the public, the media, and critics. Admittedly, this communication is difficult to do in a society that defines competence as knowing what one is doing. In the case of planning for a HLNW repository, it is time for everyone involved to admit that this repository is an extremely uncertain undertaking, more akin to an experiment than to a well-established technological exercise.

Whether willing participants in the experiment or not, the people whose lives will be affected by the uncertainties associated with the repository—and whatever consequences it produces—should be part of its planning, design, and evaluation. It is possible to develop a siting process that encourages such participation, as experience in other countries demonstrates (see chapter 6). Moreover, by frankly acknowledging the limitations of the available information and predictive techniques, program managers can actively promote social trust and elicit a readiness to tackle the problem from a different direction if necessary. For this strategy to succeed, however, ·a cautious and deliberate approach is needed, and the stages of repository development must allow time to discover whether the predictions and expectations associated with the project are reasonable and likely to inspire confidence in the future. This approach, which eschews the misguided notion that solutions to complex and unique uncertainties can be developed according to a predetermined schedule, has been adopted by almost all industrial nations except the United States.

The techniques and models used for assessing the program must be viewed primarily as learning tools, not as crystal balls that foretell the future. And it's equally important to put in place a monitoring system to alert us to what is actually happening and to improve our understanding of the realities that influence project development. The schedule should proceed in a step-by-step fashion that permits—in fact, encourages—everyone to stop and reflect and, if necessary, make mid-course corrections mandated by new technical developments and changing social values.

The key to this strategy is developing a management approach that is adaptive—that is, it must have the ability to adjust to technological and social change and unforeseen circumstances. Such an approach requires the following attributes and attitudes, as outlined by Walters (1986) in his work on adaptive policy design for complex problems:

- Action cannot wait for a full understanding of social and scientific issues that confront policy makers, particularly in the case of unique or first-time experiments like HLNW disposal.
- An explicit admission of uncertainties and ignorance is required, as is the use of imagination. It is necessary to de-emphasize the application of methods that have withstood the test of time, because they are often inappropriate, especially in confronting a challenge for which we have no previous experience.
- Misunderstandings, breakdowns, errors, and false starts are inescapable, and surprises will happen. We have no analytical tools that can completely eliminate the element of surprise so we are forced to come up with ways to cope. These ways include monitoring systems that help us learn as we go; redundancy in protective systems; technical reversibility or repairability (currently being advocated in Sweden); multiple geologic and engineered barriers; and, finally,

multiple disposal sites. Measures such as these will enhance the adaptability of the management system, reduce the possible consequences of any mistakes and mishaps that occur along the way and increase the chances of recovering—and learning—from mistakes while retaining public support and confidence.

Adaptability is only one of the characteristics needed by the organization charged with siting, building and operating a HLNW repository, however. It also requires reliability (do the job without error), durability (survive for the length of time needed to complete the job), stability (continue performing the required tasks despite external changes), capacity (handle the required volume of wastes), and integrative management (consider the system-wide consequences of decisions).

In conclusion, an approach that explicitly recognizes the experimental nature of nuclear waste management is clearly needed (Cook, Emel, & Kasperson, 1990). The National Academy of Sciences Board on Radioactive Waste Management also endorsed this type of approach. Such an approach must be adaptive, allowing adjustment to new information, advances in technology, and changing social values and environmental conditions and, in the process, instilling a sense of confidence in the public that the program can accomplish its mandate safely, fairly, efficiently, and economically. Such a program will be focused on discovery, understanding, and the ability to adapt and adjust to change. These values stand in stark contrast to current efforts to demonstrate the suitability of a politically chosen site by manipulating scientific and technological models, methods, and data as strategic tools to accomplish management goals.

The following chapters explore the issue of public trust in the HLNW program, and the extent to which the management of the current program has failed to achieve the qualities described above.

5

Public Responses to High-Level Nuclear Wastes

The public most often responds to the idea of having a high-level nuclear waste (HLNW) facility located near their communities or in their state with fear, distrust, and fierce opposition. Only a few communities—usually those historically associated with other nuclear facilities such as power plants or weapons manufacturing—have shown any willingness to host a nearby repository. Elsewhere, people find radioactive materials to be the least acceptable of hazardous wastes.

So profound has this public rejection of HLNW become that the mere association of a HLNW facility with a place is enough to generate extremely negative images in the minds of local residents and visitors, creating a form of stigma that has important social and economic consequences. This stigma reduces a community's attractiveness and property values. In a recent New Mexico case (*City of Santa Fe, New Mexico, v. Komis*, 1992), for example, the courts awarded a landowner $337,000 for losses because a highway to be built adjacent to the property will be used for HLNW shipments to the Waste Isolation Pilot Plant (WIPP) near Carlsbad, more than 250 miles away. The New Mexico State Supreme Court upheld this award, noting that "a negative public perception exists about the WIPP route" which, whether "well founded or not," constitutes a legitimate basis for damage payments to landowners. This decision has important ramifications for the waste disposal program, suggesting that significant costs may be associated with its stigmatization of communities near a repository or adjacent to HLNW transportation routes.

A HLNW repository evokes the public's fears about accidents and their potential health and environmental effects, prompts public concerns about economic harms to their communities, and raises questions about declines in their quality of life. The potential for these stigma effects should not be underestimated, particularly in a state like Nevada with an economy so dependent on tourism and gaming. Indeed, the proposed Yucca

Mountain repository can be viewed as a huge gamble with Nevada's future economic well-being. Even in a state that thrives on betting, it's a gamble Nevadans would rather not take.

These social and economic concerns are compounded by a deep and growing distrust of government institutions and the nuclear industry, and doubts about their abilities to manage large nuclear programs safely and efficiently. Their track record has not inspired confidence, nor has their past approach to dealing with public concerns and political opposition. Events like the leaks from nuclear waste containers at Hanford, Washington, as well as evidence of mismanagement at other DOE facilities, such as the weapons production site at Rocky Flats, Colorado, have shaken public confidence. Government agencies and the nuclear power industry have progressively lost support over the years, and it may be exceedingly difficult for existing players to regain the level of public trust required to succeed in siting, building, and operating a HLNW repository.

Meanwhile, admissions of flaws in risk assessment techniques have increased the public's skepticism about the often-sanguine reassurances of experts. (Some scientific and technical expertise has developed within anti-nuclear groups, but most expert opinion remains aligned with government agencies and the nuclear industry.) The admissions also reinforced the public's determination to gain greater access to and control over the decision-making and policy processes. The loss of public confidence is particularly salient to the problem of HLNW disposal, not just because HLNW are involved, but because they must be handled and stored over a long period. That this is a challenge unprecedented in human experience only intensifies the public misgivings instilled by past failures.

Opposing Models for Managing Nuclear Technologies

Much public disenchantment with the federal government's handling of nuclear issues can be traced to the fact that a single agency—the Department of Energy (DOE)—has been given two distinct and often conflicting mandates: to develop and promote nuclear technology and to ensure the public's safety from HLNW. Historically, this dichotomy bred a climate of tension and often one of outright conflict—a situation that, in the early 1970s, lead to the demise of DOE's predecessor agency, the Atomic Energy Commission (Walker, 1992). DOE's critics charge that DOE has traditionally shown a bias toward its mandate of promoting nuclear development.

Federal officials have not been very responsive to concerns expressed by the public either. These officials often seem to interpret their duty as to act on behalf of the public, but not necessarily to respond to public concerns. According to this view, the public's influence should be channeled through its elected representatives. However, politicians must necessarily depend on the scientific and technical expertise of federal agencies, the nuclear industry, universities, and research centers. Indeed, the federal government's approach to large technological ventures such as building a HLNW repository generally follows a simple model: Scientists do the research necessary to determine the nature of the problem and the challenges it presents, the technological possibilities for solving it, and the benefits and risks of proceeding with the project; engineers design and build the machines and equipment needed to realize the benefits and reduce or eliminate the risks; and, finally, management experts organize and coordinate the development and implementation of these technological solutions. According to this idealized model, nuclear power plants provide abundant and economical electric power with a minimum of health and safety risk to workers or the public. Likewise, HLNW from civilian and defense programs are handled, stored, and permanently isolated in a safe, efficient, and economical way.

Unfortunately, in the nuclear field, as in most things, reality falls short of the ideal. Nuclear technologies are among the most sophisticated and technologically demanding humanity has ever developed. They require extraordinary capital investment, scientific and technical expertise, and vigilance. And they require public support, or at least acceptance—a fact that many experts and managers of the nuclear enterprise regard primarily as an irritating complication in need of careful, even manipulative, handling. These experts generally find it extremely difficult to come to grips with the fact that the public is rarely sympathetic to a technological fix approach to nuclear issues. Many people, of course, do not have the education, professional background, or inclination to do detailed analyses of the scientific and technical data thrust on them. But that's not the real issue—although it often misleads nuclear experts into thinking that if they redouble efforts to educate opposition will melt away. However, most people who are asked (or forced) to host a nuclear facility in their backyard are—regardless of their education and background—apt to mix other more personal criteria into their evaluation, including fears for their safety, concerns for property values and quality of life, psychological factors (e.g., tolerance for risk-taking), and value judgments as to the relative weight given to risks and benefits. These judgments may differ markedly from those made by the nuclear industry.

The public also factors in how much they believe the information provided to them—and, particularly, how much they trust the messengers who gave it. The potentially catastrophic consequences of nuclear accidents and the many uncertainties associated with technical risk assessment create a public demand for a high level of demonstrated competence and trustworthiness in those mandated to operate a HLNW disposal program.

Unfortunately, government agencies have in the past resorted to secrecy and repression of information to escape public accountability for the problems at their nuclear projects. This approach has had two baneful results: when the public discovered these deceptions, the government agencies involved suffered a serious loss of trust and confidence, and practicing these deceptions created an adversarial culture within the nuclear community that tended to dismiss the legitimacy of public opinions and attitudes (U.S. Secretary of Energy, 1993). This culture developed during a time of rapid growth in nuclear technologies and increasing confidence by scientists and nuclear experts that they understood and could control the challenges of the atom. This confident attitude contrasted with that of the public, whose concerns tend to be based more on personal and social values rather than strictly scientific or technical factors. As a result, nuclear experts often came to regard public concerns as unscientific and to believe they should be accountable only to their colleagues and peers.

Nowhere was the disparity between expert confidence and public acceptability more pronounced than in the field of nuclear waste management. Yet the public's misgivings had merit. It's clear in retrospect that the experts knew considerably less than they thought they did, and, as a result, the nation now faces an unprecedented and expensive effort to clean up radioactive contamination, with unknown risks to the health and safety of workers and the public. Perhaps the most dramatic case is the DOE site at Hanford, Washington, where containers of waste from the defense program have been stored since World War II. Today, more than a third of the area (over 200 square miles of the 570-square-mile reservation) is contaminated with radioactive waste, including massive leaks from liquid HLNW storage tanks. Although many containers are known to be bad, the exact state of their contents and the amount of danger they present is not known.

Unfortunately, Hanford is not the only example; there have been failures at other DOE sites where defense wastes have been handled, including Rocky Flats, Colorado; Oak Ridge, Tennessee; Savannah, Georgia; Fernald, Ohio; Pantex, Texas. Some problems originated years ago, even decades ago. The fact that they have not been properly managed in the interim is distressing. In response, DOE points out that past waste management practices were not designed for current standards, and while this statement does not excuse the results, it is something of an explanation. What is less understandable is the December 7, 1993, report in the *New York Times* that unprocessed fuel rods in DOE pools at three sites (Hanford, Savannah River, and INEL in Idaho) are decaying, leaking, deforming, and exposing workers to sharply higher radiation levels. This is now, not then.

Commercial waste management programs have come to grief as well. A commercial reprocessing plant in West Valley, New York, all too quickly became a difficult radioactive waste problem. And a low-level nuclear waste facility at Maxey Flats, Kentucky, experienced failures in scientific evaluation, engineering design, and operations management before it was closed in 1977 for safety reasons.

By the time NWPA was passed in 1982, it was clear the public had little confidence in the competence and intentions of federal agencies responsible for disposing of the nation's HLNW. A series of failures in early attempts to locate, even to study, potential HLNW repository sites compounded problems. These failures and controversies made the development of new facilities anathema in almost all communities and states where they were proposed—attitudes that intensified during the 1980s as the long-standing problems at the nation's defense facilities were becoming increasingly obvious.

Recognizing the potential for conflict between the federal HLNW program and the states and communities proposed as repository host sites, Congress tried to move beyond the simple model of governmental duty as defined by scientists, engineers, and managers. With its focus on equity, benefit-sharing, and federal-state consultation, NWPA was, in many ways, an innovative, even revolutionary, effort to engender support for the repository program through greater public access, control, and empowerment. And this effort was not easily accomplished—it took a long drawn-out contest in Congress to achieve the idea that state and local interests should be included in repository decision making.

DOE was to implement these innovative aspects of the law. Obviously unprepared for this challenge, the department regarded proposed host states and communities not as possible allies, and certainly not as partners, but only as potential service providers for solving the nuclear industry's waste problems—and as probable adversaries.

In the end, DOE implemented NWPA according to its long-standing model. DOE would act on behalf of the public and advance the development of nuclear technology; DOE would depend on science, engineering, federal authority, and its legal mandate; it would educate and inform the public as best it could; and it would meet the timetable, if possible. But DOE had a special regard for the interests of the nuclear power industry, which was perceived as its ally in developing nuclear technologies. DOE also had a client allegiance to the nuclear utilities, which were collecting fees for the Nuclear Waste Fund that financed the repository program. In terms of its fiduciary responsibilities, therefore, DOE viewed the nuclear industry as its client. Other parties, such as states and communities, were viewed as stakeholders if they appeared to have a creditable interest in DOE's activities.

The model inferred by the public and public advocates, including potential host states and communities, is quite different. Technical competence is expected and evidence of scientific failing or dissembling on research findings is fatal to public trust and confidence. Beyond this expectation, the public demands candor and fiduciary responsibility from its agencies, perhaps even more than from elected officials. In the case of a HLNW repository, this expectation means absolute commitment to public health and safety and the environment. The confusion that DOE exhibits about who it represents—the public and its communities or the nuclear industry and its rate base—creates only suspicion and cynicism about the risk messages coming from repository advocates within government agencies. The long history of DOE service to its objectives and those of the nuclear industry, often at the expense of the public and even DOE workers, exhibits the basic conflict between agency culture and public values that has often been remarked about, even by the highest DOE officials who have come to their offices with the intention of setting things right.

Studies of Perceptions and Social Behaviors

The intense and sustained public opposition to HLNW facilities creates the possibility that such facilities will produce special adverse socioeconomic impacts. Starting in 1986, research on these potential impacts was conducted by a study team funded by Nevada's Nuclear Waste Project Office (NWPO). These studies focused on the public's perception of the HLNW disposal program, how this perception affected attitudes and social behaviors, and the potential consequences for the social and economic well-being of communities near HLNW facilities and transportation routes. Specifically, the studies dealt with the potential impact of the Yucca Mountain HLNW repository on the welfare of Nevada in particular, and examined the broader question of how society is to develop, control, and benefit from scientific and technological enterprises that also pose risks to workers and/or the public.

One significant result of this work was to provide important insights into how large technological ventures like a HLNW repository can create negative stigma effects that can, by their mere existence, seriously harm a community even in if nothing goes wrong with the facility. Stigma, which can apply to people, products, places, or technologies, is not merely a negative judgment or a simple preference for one thing over another. The stigmatized object is judged to be extremely undesirable, dangerous, or repugnant; something to be shunned and avoided. A stigmatized technology is often hazardous, but it also offends or threatens intensely held social, moral, and cultural values.

Study Objectives

The studies of public perceptions and social behaviors had two main goals: (1) to document the strength of public opposition to the repository program and concerns about its risks and potential stigma effects among residents of Nevada, and (2) to explore how public opinions and attitudes about a HLNW repository at Yucca Mountain might affect the willingness of people outside Nevada to vacation in the state, to move there for employment or retirement, or to invest there. The objectives were to understand why the public responds as it does to nuclear waste disposal projects—specifically the Yucca Mountain repository—and to determine the behavioral outcomes that might result from their perceptions of the risks involved. These goals are not, in and of themselves, anti-repository. They deal with the most basic concerns that people have with a HLNW facility.

Baseline Opinions and Attitudes

More than 20 surveys were conducted between 1986 and 1993 to assess public attitudes and opinions about the repository and its management. A description of the study team reports, including each survey, is contained in the Yucca Mountain socioeconomic study team report (1993). Concurrent surveys were conducted in Nevada, southern California, Phoenix, Arizona (major sources of visitors to Nevada), and nationwide. The most extensive surveys were conducted by telephone in the fall of 1989. More than 2500 respondents were asked about their perceptions of the risks and benefits associated with a HLNW repository, their support or opposition for the DOE repository program, their trust in DOE's ability to manage the program, and their views on other issues pertaining to radioactive waste disposal.

Respondents were asked how close they would be willing to live to 10 kinds of facilities. An underground HLNW repository was the most undesirable by a considerable margin, the median acceptable distance (200 miles) being twice that for the next most undesirable facility, a chemical-waste landfill, and three to eight times the acceptable distances from oil refineries, nuclear power plants, and pesticide manufacturing plants.

More than 75% of Nevada respondents also believed that highway and rail accidents will occur in transporting HLNW to the repository, and the response in the Californian and nationwide surveys was only slightly lower. People expressed similar expectations of problems resulting from future earthquake or volcanic activity at the site, contamination of underground water supplies, and accidents while handling the material during burial operations.

Two-thirds of the southern California respondents and three-quarters of the national respondents said a state that does not produce HLNW should not be forced to serve as a site for a repository. (The Nevada survey did not ask this question.) And a majority of those polled in southern California and nationwide judged a single national repository to be the least fair disposal option. The other options were storage at each nuclear plant, in each state, or in regional repositories, such as one for the east and one for the west.

Respondents also revealed a strong distrust of DOE. Three-quarters of Nevadans surveyed, and two-thirds of respondents in southern California and nationwide surveys, expressed doubts that DOE would promptly and fully disclose information about accidents or serious problems with the waste management program.

These data demonstrate that the public has a strong aversion to hazards posed by nuclear waste facilities and confirm findings from many other sources, including surveys done by other researchers, literature reviews, focus groups, experimental studies, participant/observation studies, and examination of analogous cases (e.g., other proposed HLNW sites, low-level radioactive sites, nuclear power stations, and defense production facilities). In fact, the public considers a HLNW storage facility to be more dangerous than nuclear power plants or weapons production facilities, while many experts think geological storage is the least risky of these enterprises.

In summary, United States citizens could hardly have a more negative opinion of HLNW disposal; as studies discussed in more detail below demonstrate, HLNW are perceived as representing very high risks, which the public characterizes as immense, dreaded, catastrophic, unknown, uncontrollable, and inequitable.

Support and Opposition in Nevada

In the 1989 survey, nearly 70% of Nevada residents said they would vote against a repository at Yucca Mountain, compared with about 14% who said they would vote for it. About 68% of Nevadans strongly agreed that the state "should do all it can to stop the repository" and another 12.5% agreed somewhat with this statement, while only 16% disagreed. Three-quarters were in favor of Assembly Bill 222, which made it illegal to dispose of HLNW in Nevada; only 18.4% were opposed. Finally, nearly 75% of Nevadans said the state should continue opposing the repository even if this opposition means turning down benefits offered by the federal government; less than 20% said the state should stop fighting and make a deal.[1]

Follow-up surveys in Nevada in October 1990, March 1991, November 1991, March 1993, and September 1994 confirm that high levels of opposition and distrust persist. The percent of Nevadans who may vote against a Yucca Mountain repository has remained in the 70% range, and opposition continues to outstrip support by margins of three or four to one.

The surveys reveal that opposition to the repository in Nevada is based on concerns about potential hazards of HLNW, but the opposition also reflects a profound lack of trust in the federal government—a distrust that embraces not only DOE, but Congress, the Nuclear Regulatory Commission (NRC), and even the Environmental Protection Agency (EPA). Within Nevada deep resentment exists about the unfairness and inequity of the HLNW program, especially the 1987 amendments to NWPA, which targeted Yucca Mountain as the sole potential repository site. Nevadans felt very strongly that the state should be able to decide whether to accept the repository and that state and local officials should be genuinely involved in decisions about Yucca Mountain.

Finally, concern about the potential impacts of stigmatization on the tourist and visitor industries, which dominate the state's economy and revenue base, was evident. Many state officials believe Nevada's economy could be vulnerable to the stigma effects of a repository in a way no other state probably would be, because Nevada is the only state so dependent on attracting visitors for gaming and entertainment.

So deep-rooted is opposition to the repository in Nevada that a recent nuclear industry advertising campaign (sponsored by ANEC) was greeted by officials, citizens, and businesses with disbelief and ridicule. Despite the millions of dollars spent by the industry on advertising and intensive lobbying, in 1993 Nevada's legislators rejected (by a margin of two to one) a resolution calling for a study of whether the state should negotiate for benefits while site characterization is going on. This development underscored the intensity and persistence of the political opposition within the state. DOE and the nuclear industry now view state opposition as a serious obstacle that may prove decisive in the future. Cost overruns, schedule delays, other management failures, and the continuing weight of Nevada's active and unrelenting opposition cannot help but compound difficulties the federal government and nuclear industry are having in siting a permanent HLNW repository.

---

1. These figures do not include respondents who did not answer certain questions or those who had no opinion. Therefore, these percentages may not total 100%. Readers interested in the details of the surveys are encouraged to contact any of the authors for documentation reports.

Stigma and the Nevada Economy

Nevada's economy is dominated by the tourist and visitor industries and is unique in its dependence on gambling and entertainment. Nowhere is the structure of this economic base more obvious in the Las Vegas metropolitan area, which contains about two-thirds of the state's population. Visitor spending provides employment, income, and sales, as well as state and local public revenues, which are structured to benefit from nonresidents. Each year more than 20 million visitors come to Las Vegas and spend well over $10 billion. Growth in this sector of the economy has fueled the building of new hotels, casinos, and convention and entertainment facilities. Much of the growth of the Las Vegas visitor economy has been based on an effort to widen the entertainment base beyond gaming, and to attract a more diverse visitor base including families, business meetings, and conventions. Other areas of the state (e.g., Reno, Lake Tahoe) also depend on tourists and visitors, although these areas do not generate income and revenues comparable to that of Las Vegas.

A key issue explored in the socioeconomic studies, therefore, concerned the potential for a Yucca Mountain repository to create stigma effects that might undermine the economic base of southern Nevada and that of the state as a whole.

One approach was to ask survey respondents to indicate whether having a repository located 100 miles away (approximately the distance between Yucca Mountain and Las Vegas) would reduce a community's desirability as a place to attend a convention, take a vacation, raise a family, retire, or locate a new business. In the 1989 national survey, nearly 75% of respondents said a repository would reduce the desirability of the community for raising a family, while 41% said a repository would reduce their willingness to attend a convention. Another study, done by University of Colorado researchers, found that job seekers could be deterred from migrating to Las Vegas by the presence of a repository at Yucca Mountain and by reduced employment opportunities if tourism decreased (Greenwood, McClelland, & Schulze, 1988).

A survey of convention planners conducted by the Wharton School, University of Pennsylvania, evoked particularly adverse reactions. The planners were asked to comment on seven scenarios associated with the repository over a 10-year period, starting with a benign history with no problems and escalating through multiple mishaps. Each case was divided into two subsets, one in which problems and accidents receive only brief media attention and another in which these events are the focus of intense local and national coverage. For all scenarios, nearly a third of the planners ranked Las Vegas lower as a meeting site, compared with their original rankings. In the extreme case—a scenario with recurrent accidents and high media attention—75% of the respondents lowered their ratings, and nearly 50% said they would no longer consider Las Vegas as a candidate site for a convention.

A 1988 survey of 400 members of the National Association of Corporate Real Estate Executives revealed that, like members of the general public, they felt the existence of a repository within 100 miles of a community would detract from its suitability as a location for administrative offices, business and professional services, and businesses to serve the hospitality industry (Mountain West, 1989).

Forecasting Behavior Using Imagery

In surveys, experiments, focus groups, and interviews, the general public and decision makers in special areas reported that a HLNW repository at Yucca Mountain would influence their decisions and actions in ways that could negatively affect Nevada's economy. Nevertheless, these studies are limited in their ability to forecast actual future behavior, especially with regard to a unique facility with which no one has had any experience, and in response to events that will take place years, decades, and perhaps even centuries from now. Therefore, another set of studies was developed, based on the notion of environmental imagery, to test three propositions:

1. Images have been associated with environments and places, and these images have diverse positive and negative effects on people's preferences about places they wish to vacation, retire, raise a family, or invest.
2. A HLNW repository evokes many strongly negative images, consistent with perceptions of extreme risk and stigmatization.
3. The Yucca Mountain repository and the negative images it evokes will, over time, become increasingly salient in the images of Nevada and Las Vegas.

If these three propositions are true, the attractiveness of Las Vegas and Nevada would decrease if a HLNW repository is sited at Yucca Mountain. This decreased attractiveness will result in negative economic impacts on the state and its communities. These impacts could become doubly significant in the future, when Nevada can expect stiffer competition from the growth of casino gambling in other states.

The three propositions, if true, suggest a mechanism by which a HLNW repository could adversely affect a vulnerable community. This mechanism can be demonstrated without asking people to predict future behavior. The methods and results of these studies have been published elsewhere (Desvousges, Kunreuther, Slovic, & Rosa, 1993; Slovic, Layman, Kraus, Chalmers, & Gesell, 1991), but a summary is provided below.

The first task was to elicit images from survey respondents. In a telephone interview, respondents were asked to think about a stimulus (e.g., "Las Vegas" or "Nevada") and to say the first thought or image that came to mind. Up to six images were elicited from each respondent, and the process was repeated for other cities and states. This process was also used to obtain images for "underground nuclear waste storage facility." Respondents were asked to rate each image on a five-point scale from very positive to very negative; then they were asked to provide preference ratings related to behaviors (e.g., visiting, migrating to, investing in), for the stimuli.

Images and preference ratings for cities and states were strongly related—that is, the more positive the images a city or state elicited, the more likely that place was to be preferred over other cities or states. Similarly, responses obtained from corporate real estate decision-makers revealed a similar ability to predict their preferred locations for siting a new business based on their images. In addition, people who had positive images of gambling were influenced as much by other positive and negative images as people who were less approving of gambling.

Some Phoenix, Arizona, respondents were re-interviewed 16 to 18 months after the initial interviews in 1988. Their earlier image scores for different places significantly increased the ability to predict their vacation behavior between 1988 and 1989. This finding supports the first proposition that images are associated with cities and states, that these images define positive and negative meanings for places, and that these meanings influence behaviors regarding places.

In the Phoenix surveys, about 10% of the responses to "Nevada" contained nuclear imagery, including terms like nuclear testing, nuclear bomb, nukes, explosions, and radiation. These respondents gave Nevada lower image ratings and preference rankings than those who did not associate the state with things nuclear—evidence that a nuclear image could have a negative impact on preferences for places.

Additional surveys examined images applied to underground HLNW repository. Some 10,000 images were elicited, and more than half fell into four categories that were dominated by the words dangerous, danger, death, and pollution. Other image categories reflect distrust and concern about mistakes and accidents or were visceral negative responses such as stupid, illogical, terrible, ugly, disgusting, evil, and insane.

Positive images were rare. Concepts such as necessary, employment, and money/income comprised only 2.5% of the images, while the image safe was below 0.5%. In total, less than 5% of the image categories were positive, accepting, or neutral. Surprisingly, almost no positive images associated with energy and its benefits—for example, electricity, light, heat, employment, health, progress—were observed. Nor did the concept of a waste repository or storage facility elicit many images that reflected the view of many experts that such facilities are safe and necessary.

These findings substantiate the second proposition—that a HLNW storage facility evokes consistent, extreme, negative images. The modest associations of Nevada with nuclear imagery because of the weapons test site provided indirect evidence for the third proposition—that a HLNW facility will also become associated with the public perception of Nevada if the Yucca Mountain project proceeds.

These images, as well as responses to attitude and opinion questions, reveal a powerful aversion to nuclear waste, characterized by pervasive feelings of dread, revulsion, and anger—the raw materials of stigmatization and political opposition.

Origins of Public Responses to Nuclear Technologies

Images and attitudes so negative and impervious to opinions of technical experts must have potent origins. Weart's (1988) historical analysis showed that nuclear fears are deeply rooted in our social and cultural consciousness. Other studies (Fiske, Pratto, & Pavelchak, 1983; Slovic, Lichtenstein, & Fischhoff, 1979) have shown that concepts like nuclear wars and nuclear reactor accidents evoke images similar to those evoked by the concept of a HLNW repository. The shared imagery of nuclear weapons, nuclear power, and nuclear waste may explain why a HLNW repository is judged by the public to pose risks at least as great as a nuclear power plant or a nuclear weapons test site.

Further insights into the special quality of nuclear fear are provided by Erikson (1990), who describes the exceptionally dread quality of accidents that expose people to radiation. Unlike natural disasters, these accidents have no end. "Invisible contaminants remain a part of the surroundings—absorbed into the grain of the landscape, the tissues of the body and, worst of all, into the genetic material of the survivors. An all clear is never sounded. The book of accounts is never closed" (p. 118).

Perceptions of extreme risk from HLNW, deeply rooted psychological and cultural views, experiences of failed waste management, lack of trust in the abilities and intentions of the government and nuclear industry, concerns about social and economic impacts that will result from the stigmatization of a repository location, and the sense that the people who will be most affected by the repository have no control over its development and operation combine to create a potent public opposition to nuclear waste facilities.

Communicating Stigma

The effect of a HLNW repository in stigmatizing adjacent communities or an entire state would be greatly influenced by how much people in those communities—and elsewhere in the state and the country—know about the repository itself and, especially, about any incidents or accidents that occur. Similar effects can materialize even before the repository is built, if negative information is disseminated about a proposed host site.

Such information will be widely retailed to the public, especially through the news media. In part, this high public profile exists because the nuclear waste program has national roots—only 14 states (the 10 western states include North Dakota, South Dakota, Montana, Idaho, Wyoming, Oklahoma, Colorado, New Mexico, Utah, and Nevada, and the 4 eastern states include Rhode Island, Delaware, West Virginia, and Kentucky) do not have nuclear power stations where spent fuel rods are stored on site. Moreover, decades of failed attempts to locate a repository have generated thousands of news stories about HLNW hazards. The actual development of a repository can be expected to attract equally high levels of media attention, especially if problems or accidents occur.

The transportation of HLNW from reactors and government defense facilities, which requires thousands of shipments over at least three decades, will spread the direct impacts to additional states and communities along highway and rail routes. This transportation effort will also effect future generations. The cost of locating, building, and operating just the first repository is immense—currently estimated at $35 billion. The program's budget and progress are reviewed each year by Congress and are closely watched and reported by the national, regional, and local media. The connection between the repository and the nation's weapons facilities further adds to the intrinsic newsworthiness of nuclear waste issues.

It should not be surprising, therefore, that the media report extensively on issues related to management of the nuclear program and that the public pays close attention. This coverage sets the stage for a phenomenon known as the social amplification of risk.

Social Amplification of Risk

Early in the Nevada socioeconomic research project, the study team was intrigued by the fact that some risks (i.e., evaluations or perceptions of a hazard) or risk events (i.e., incidents that effect evaluations or perceptions of risks) regarded as relatively minor by technical experts often provoke much stronger concern among the public, resulting in substantial social and economic impacts. This phenomenon was first reported in a theoretical paper on the social amplification of risk (Kasperson, Ratick, &

Renn, 1988) and since then, more than a dozen publications have addressed the issue (for a review of this literature see Kasperson, 1992).

According to the theory, the social amplification of risk involves two major stages (or amplifiers)—the transfer of information about the risk or risk event and the response mechanisms of society. Many people do not experience the risk directly, but do experience it through media coverage, where these two amplifiers are combined, often in an iterative fashion. When this combination happens, as in the case of the Three Mile Island accident, the resulting social amplification comes from the volume of news, the degree of dispute about what has happened and why, the extent of dramatization that makes the news memorable, and the symbolic or signal quality of the information. One of the most influential signals has to do with the impressions people receive concerning the ability of experts to manage and control the hazards of a technology.

A highly amplified risk event or incident can trigger the negative imagery associated with undesirable and hazardous places. As Kasperson, Ratick, and Renn (1988) state, "since the typical response to stigmatized... environments is avoidance, it is reasonable to assume that risk-induced stigma may have significant social and policy consequences" (p. 186). For Nevada, these consequences could include the possibility of a negative economic impact, should this avoidance reaction occur in people who would otherwise visit, move to, or invest in the state.

A Crisis of Trust and Confidence

The public's fear of and revulsion from a HLNW repository stands in contrast to the confidence that most technical analysts and engineers have in their ability to dispose of radioactive materials safely. For example, a National Research Council/National Academy of Sciences (1990) report was highly concerned about the difficulties of predicting the long-term performance of a repository, but concluded that "these uncertainties do not necessarily mean that the risks are significant, nor that the public should reject efforts to site the repository" (p. 3).

Many risk analysts have argued that public acceptance of any risk is more dependent on trust and confidence in risk management than on quantitative risk estimates. The gap between confident views of experts and skeptical views of the public regarding HLNW disposal suggests a serious crisis of confidence and a profound breakdown of trust in scientific, governmental, and industrial managers of nuclear technologies. This breakdown was clearly evident at the time NWPA was enacted and has been documented repeatedly in subsequent public opinion surveys. NWPA required DOE to analyze alternative ways to manage its civilian radioactive waste facilities [Section 303]. An advisory panel was established; it submitted its report in December 1984. The panel recommended that a dedicated, federally-chartered corporation be created to replace DOE because DOE lacked the credibility, management flexibility, and cost-effective administrative capabilities to manage the waste program. But the U.S. Secretary of Energy never acted on this recommendation.

Concerns about lack of trust and confidence in DOE have been expressed in reports by the United States General Accounting Office (GAO), the Office of Technology Assessment (United States Congress; OTA), and the United States Nuclear Waste Technical Review Board, as well as by many social scientists, policy analysts, and state and local governmental officials. Eventually, in 1991 the U.S. Secretary of Energy created an Advisory Board Task Force on Radioactive Waste Management to make recommendations on how DOE could strengthen public trust and confidence in the civilian radioactive waste management program. In their 1993 draft report, this task force reviewed the difficult situation faced by DOE and documented the profound lack of trust in the agency. The task force also made several recommendations to increase DOE's standing and reputation. However, DOE's trustworthiness may be beyond repair, a distinct possibility that the task force was not prepared to investigate. In any case, the task force suggestions, good as they may be, do not appear to be equal to the job of reclaiming public confidence in DOE, especially as DOE releases formerly secret files and information about their cold war nuclear weapons programs. For example, in 1993 the federal government reported that during the Cold War DOE had conducted 204 secret underground nuclear tests, and that more than 600 individuals had been used in experiments, including the injection of 18 people with plutonium in the 1940s. News media reports quote U.S. Secretary of Energy Hazel O'Leary as saying she "has been 'shocked and amazed' at shortcomings in the management of the nation's nuclear weapons complex" (Healy, 1993, p. A1).

Another option is to replace DOE as the agency in charge of the HLNW program. Although the task force did not consider this a practical alternative, many observers have suggested this route as a sensible option (however much it displeases DOE) because such a change addresses a central problem of the HLNW issue: public distrust in risk management. Social psychology research clearly demonstrates that trust is quickly lost and slowly, if ever, regained—a problem that will continue to plague DOE as it struggles to place its waste management efforts on track. For example, in 1989 DOE attempted to regain the confidence of the public, Congress, and the nuclear industry by rearranging its organizational chart and promising to do a better job of risk management and science in the future, but this effort was naive. So, too, was the idea that a task force could come up with recommendations that could reverse the effects of a long history of mismanagement and mistakes that precipitated distrust in the past or overcome the DOE culture and mindset that exacerbated the problem. Trust, once lost, cannot be so easily restored.

It was also simplistic of DOE officials and other nuclear industry leaders to suggest that they could change the public's perceptions and gain support by allowing people to see first-hand the safety of nuclear waste management. We are dealing here with a situation that involves incidents or accidents that may have a low probability of actually happening, but very serious consequences if they do. A facility must operate for a very long time free of damaging events before people will be convinced that the facility is safe or represents only a negligible risk. Any adverse events that occur, even relatively minor ones, will destroy a painstakingly-earned perception of safety and reinforce the perception of danger. The intense scrutiny given to nuclear power and nuclear waste issues by the media ensures that a steady stream of problems occurring all over the world will be brought to the public's attention, continually eroding trust.

In conclusion, acceptance of a HLNW facility must come not only from the American public as a whole but also from the states and communities most affected by it. The social and economic impacts of a repository must be considered as seriously as the physical, scientific, and engineering aspects of the project. National policy makers must recognize that local and state acceptance and support is a legitimate requirement for any waste management program or project.

Task forces, policy analysts, government officials, industry experts, even DOE, have repeatedly emphasized that public support and acceptance are essential for the success of nuclear waste programs. Despite this chorus of agreement, the federal agencies designated to site, construct, operate, and regulate the repository program have made little effort to understand public attitudes and opinions. The fact is that the major effort to gain this understanding of the repository program was sponsored by the state of Nevada, which opposes the Yucca Mountain project, and not by DOE or any federal agency. DOE response has been not to address these basic issues but to attempt to portray Nevada as obstructionist and to encourage nuclear industry advocates in Congress to limit the state's ability to do research by cutting funds for its oversight program. Another choice for DOE and Congress would be to address the basic public issues and concerns if they can.

In our opinion, the solution to the problem of public support for nuclear waste programs is not to repeat the past—notably DOE's failed efforts to implement principles of NWPA and the congressional endorsement of that failure with NWPAA in 1987. The solution is much more likely to lie in efforts to understand the nature of the public concerns, to establish and maintain new institutions that will be worthy of trust, and to find ways to fulfill the expectations and demands of the public rather than attempting to ignore, avoid, or override these demands.

6

International Experiences

The United States is not alone in its struggle to find a solution to the problem of high-level nuclear waste (HLNW) disposal. Other industrial countries are grappling with similar challenges and facing the same issues the United States program has experienced. These include public antipathy toward living near a HLNW site, distrust of the experts managing HLNW, outright skepticism about experts' technically-based reassurances, fears about health effects and potential accidents, and intense political opposition.

In short, the waste disposal task has turned out to be very different than originally anticipated, and outside the experience of those mandated to find solutions. An undertaking that many in the nuclear business had viewed as a purely scientific and technical exercise has instead taken on complex social and political dimensions. The institutions charged with waste management and disposal have come under close public and media scrutiny and are confronted with a host of novel issues: protection of future generations, value conflicts, equity among regions and communities, limits of scientific knowledge, and trustworthiness of science and scientists.

The problem is compounded by the fact that a federal policy for a disposal process was not placed into law until 1982 with passage of the Nuclear Waste Policy Act (NWPA)—decades after the nuclear age started and long after wastes began to accumulate. This delay may have resulted, in part, from the notion that plenty of time existed in which to find a solution to the waste problem. This ill-founded assumption failed to foresee or acknowledge the many social and political impediments that would delay disposal and confound its managers.

Among the countries grappling with the problem, there is a consensus about the preferred technical approach—deep underground or geologic disposal. But notable differences also exist, mostly related to differing political cultures and governmental systems. It is increasingly apparent that many countries are adapting more rapidly and with more flexibility to these HLNW disposal challenges than the United States (GAO, 1994). In this chapter we examine these approaches and compare them with the United States' program, which stands almost alone in many basic elements of its management strategy and its seeming inability to respond constructively to social and political challenges.

Waste Management in Five Countries

The United Kingdom

In many ways, the United Kingdom's experience with HLNW parallels that of the United States. Their experience shows how dramatically wrong things can go, but also reveals some promising changes. The nuclear waste program in the United Kingdom involves many different companies and government agencies and has historically been intimately linked with the reprocessing of nuclear wastes. In the late 1970s, the management approach was to assign the disposal task to an existing agency (the United Kingdom Atomic Energy Authority), to conceive the task as basically a technical matter, and to conduct the exercise in relative secrecy.

In 1981, after meeting with stiff and widespread public resistance, the government canceled the disposal program. More than a decade later, there is still no active effort to site a repository; the small remaining research program focuses on the applicability of work done in other countries (Hardin & Ahagen, 1992). The intention now is to vitrify and store HLNW for at least 40 years in an interim facility under construction at the reprocessing plant in Sellafield. The development of dry storage technology has given greater prominence to the idea of deferring reprocessing and long-term HLNW disposal. The program aimed at disposing of low- and intermediate-level radioactive wastes also encountered difficulties and was restructured three times during the 1980s.

But there are signs of change. In 1982 a new agency, the Nuclear Industry Radioactive Waste Executive (NIREX), was created, and by 1987 the traditional decide-announce-and-defend siting strategy was scrapped in favor of a new approach called "The Way Forward." This more open process involved presenting different technological options for the public to consider. NIREX also embraced a new siting principle: Areas without prior experience with or relative public acceptance of nuclear activities should be avoided.

More than 60,000 copies of "The Way Forward" document were distributed for public comment. Six months was allocated for public meetings, and a British university was engaged to track public responses and petitions. While it is too soon to judge the outcome of this approach, it is apparent that a redefinition of the waste problem and a shift to a new way of doing things is in progress (Blowers, Lowry, & Solomon, 1991; Openshaw, Carver, & Fernie, 1989).

France

The changes in France are even more striking. More than any other country, France has been a nuclear state, with 51 operating nuclear power plants producing more than 70% of the nation's electricity. The comparable figure in the United States is 22%.

Since the 1973 oil crisis, the French government has been committed to increasing national energy independence by substituting uranium for oil. Electricité de France (EDF), the government-owned electric power monopoly, is the sole customer for domestic nuclear power plants. France has aggressively pursued the development of a full nuclear fuel production and reprocessing cycle, and its commitment to reprocessing and waste-vitrification has made France the world leader in developing reprocessing and related waste-handling facilities. All spent fuel rods from domestic reactors are reprocessed, and the plutonium and uranium thus recovered are recycled. France also has contracts with a number of other countries to reprocess fuel.

As for disposing of HLNW, the French government has considerable discretion and flexibility. At present, France is investigating four sites with the intention of selecting two in which to build underground laboratories or mini-repositories for conducting experiments with emplacing wastes to see how geological structures respond. The plan is to develop one lab into a full-fledged repository. This strategy gives the French waste disposal program great flexibility and resilience; if one site turns out to be unsuitable, France has a second already up and running. If both sites are acceptable, France has a choice for the permanent repository.

The goal is to select test sites by the turn of the century, although there's no precise deadline. No timetable has been announced for the development of a permanent repository, which probably would not happen for 15 to 20 years after the test labs start up.

Unlike the situation in the United States, the site selection process is not governed by strict quantitative criteria or regulations. What's best—or even acceptable—remains an open issue to be decided by experts using their best professional judgment. Also open is the question of when to commit to geologic disposal versus long-term interim storage. Finally, the design and development of the repository have not been predetermined; they will depend on the geology of the selected site and many other variables, including costs and how long the wastes need to cool down (Cook, Emel, & Kasperson, 1991/92).

For a long time France pursued a hardball strategy in which policy issues and safety were considered to be a problem of techniques—in other words, a strictly technical matter, not a political one—and site selection a matter for experts. France has a powerful, centralized government, and its bureaucracy has a strong tradition of social engineering accomplished by the grand corps—organizations of technical and administrative experts resembling fraternal societies. Local governments and the courts are essentially appendages of the civil administration. Accordingly, the French waste management strategy has been in the hands of technocrats and firmly oriented toward the problem of techniques (Cook, Emel, & Kasperson, 1991/92).

Initially, the French government freely employed a strong arm against local resistance to nuclear projects. The Super Phoenix reactor site at Malville was surrounded by electric barbed wire and protected by dogs. In 1977, a demonstrator was killed by police at another site. And in 1980, when the mayors of Golfech and Auvillar, two towns near a proposed nuclear plant, refused to house the environmental assessment documents for the project, the state brought in trailers for storing the documents and police gassed protesters.

In the past, most national politicians deferred to experts on nuclear questions, and the experts in turn have alternately placated or overridden public concerns. But these concerns continued to mount throughout the 1980s, culminating in the coalescence of regional opposition groups into a national anti-nuclear campaign. Finally recognizing that any solution ultimately must be politically sound and workable, the government put a temporary moratorium on the disposal program in 1990 and restricted the number of sites to be considered.

Although a national debate promised by President François Mitterand has not occurred, a nuclear waste negotiator has been appointed to work with potential host communities and find prospective sites for the two testing labs. Only after lengthy characterization and assessment will a recommendation be made to the French parliament, which will then make a political decision on the disposal program and legislate the key features of repository design. In 1991 France also passed new legislation separating the National Agency for Management of Nuclear Wastes (ANDRA) from the French Atomic Energy Commission, thus increasing its independence from the nuclear industry. France also created a new national commission of independent experts that reports directly to parliament.

Germany

Germany has 21 operating nuclear plants, which produce about 40% of the country's electricity. (Five nuclear plants in the former East Germany have been shut down for safety reasons.) As in the United States, the German HLNW program has been beset by public concerns, intergovernmental conflicts, legal challenges, and political delays. The original plan was to build a large complex in which several elements of the nuclear fuel cycle—reprocessing, fuel fabrication, preparing wastes for disposal, and permanent disposal—would be concentrated in one location. The reprocessing plant within the complex was to have sufficient capacity to allow the operation of about 50 nuclear power plants (like France, Germany reacted to the 1973 oil crisis by taking the view that expansion of the nuclear option was the only way to secure future energy supplies). This new nuclear complex was to be financed and operated by private corporations, except for the permanent disposal facility, which, although financed by fees charged to the nuclear plant operators, would be operated by the government.

Responsibility for the waste program was assigned to a complex intergovernmental and management structure. Political power in Germany is quite decentralized; individual states, or *Lander*, have considerable independence and, in fact, it is the state government that would actually license a HLNW repository. This decentralization of power permits substantial local influence over policy formation and implementation. Similarly, German's professional, merit-based civil service is highly decentralized; few civil servants are employed directly by the federal government, and the central ministries tend to be small planning organizations that develop policies reliant on implementation by state governments. German bureaucrats also take a legalistic approach to administration; most high ranking public servants have law degrees and see their job as one of strictly applying the law to particular circumstances rather than making policy. As with membership in the French grand corps, civil servants in Germany have considerable social status.

This political and managerial context allows political opposition to influence policy via many different routes, and so it has been with the nuclear waste issue. When the site of Gorleben was chosen in 1979 for the large waste-processing and disposal complex, opposition in the host state of Lower Saxony made it clear that such an ambitious plan of waste concentration was not publicly acceptable. A more limited plan was pursued until an accident at the site in 1987 revitalized local opposition and led to a temporary moratorium. An unexpected incursion of briny water in the repository shaft led to another halt in 1991. Meanwhile, Lower Saxony has made a determined effort to halt repository investigations, including stopping trucks carrying wastes.

Because Germany, like the United States, is pursuing a strategy of investigating only a single site, progress on repository development has periodically halted. Meanwhile, domestic reprocessing of nuclear wastes has been abandoned because of economics and political opposition. In addition, a planned repository for intermediate- and low-level wastes at the Konrad iron mine has been delayed by legal and political challenges over the past 15 years; for example, 290,000 people from surrounding towns recently filed written objections to development proposals for the facility (Toro, 1992).

In the face of intense public opposition and continuing resistance from the repository host regions, Germany has yet to create a more resilient program or to adapt its institutions in the way France and the United Kingdom have. Indeed, the idea of transferring responsibility for waste disposal from the federal government to private industry is moot, perhaps signalling a political retreat from the waste disposal problem and a potential new round of uncertainties in Germany.

Sweden

Among advanced industrial societies, Sweden stands out for its accomplishments in managing nuclear wastes. Since 1985 it has operated an interim spent-fuel storage facility adjacent to the Oskarshamn nuclear plant south of Stockholm. By 1988, an underground disposal facility for low- and intermediate-level nuclear waste had been sited and built on the east coast and put into operation. A waste transportation system, including a specially-designed ship for carrying HLNW, has been developed, and Sweden is making steady progress in finding a site for permanent disposal of HLNW. Sweden is, in short, close to completing an effective waste management system for its 12 operating nuclear power plants. And Sweden has accomplished all this in a country noted for the strong environmental values of its populace, a tradition of local control that includes community veto rights, and a well-developed anti-nuclear movement (Ahlström, 1994; Holmberg, 1991).

What explains Sweden's success? Certainly not the magnitude of resources and personnel committed to the effort; Sweden's program is very small compared with that of the United States. A special company (Kärnbränslesäkerhet; KBS) created by Sweden's four utilities in 1972 to manage HLNW has only 70 employees, and another 250 contractors and consultants are engaged in research and development. In 1991 the program's operating cost was just US$125 million (SKB, 1992). SKN, the Swedish government agency responsible for evaluating and supervising the nuclear industry's research and development program, has 10 people who work together on one floor (NWTRB, 1991). Indeed, Sweden's single-mission institution and its small, highly integrated management system may be the key to its effective progress.

But Sweden also recognized the special qualities of the HLNW problem and based its program from the outset on two underlying moral and equity principles: (1) the repository should be designed so that surveillance and maintenance are not necessary for safe functioning, but future intervention and corrective measures are possible if the design proves to be unsuitable; and (2) responsibility for the permanent repository should not be placed on future generations, but neither should future generations be denied an opportunity to assume responsibility (SKB, 1989). Similarly, no financial burden should be passed on to future generations.

Even the industry's public information program is founded on explicit recognition of social responsibilities and contains provisions to prevent self-serving tactics (Söderberg, 1989). Both SKB and SKN operate on the philosophy that controversies cannot be settled as if they are the result of misinformation—that is assuming that people simply lack the correct information and will be converted once the information is given. Instead, they work to maintain a two-way flow of communication and take the view that it's not enough just to disseminate information to the public. They must also pay attention to what the public wants and accepts. This philosophy stems from a deep-rooted view that, ultimately, the public must decide what's to be done with HLNW.

Consequently, a thorough and well-established mechanism called the remiss process is used to actively involve the public and critics in the program. First, a proposed public policy is widely disseminated to diverse stakeholders, who evaluate and comment on the proposal in writing. The proposal's sponsors must then respond in writing to all the comments and often make modifications to encourage national consensus. Extraordinary efforts are made to encourage this dialogue, including paying for all postage on inquiry letters; maintaining a 24-hour toll-free telephone information number; using an interactive educational computer program; developing a trailer exhibition that travels throughout the country; and, finally, sponsoring exhibitions, print and televised information programs, and public visits to the waste transport ship.

In addition, the HLNW program has undergone continuing scrutiny by experts from around the world, making the program the most extensively reviewed in the world.

The Swedish approach reflects a determination to accommodate rather than override local concerns and to fashion a national consensus on HLNW disposal within which siting efforts can succeed. And perhaps most importantly, Swedish program managers give safety concerns priority over cost-benefit calculations, an approach exemplified by the proposed use of expensive but long-lasting copper canisters to contain the wastes. Such steps have helped build public support. Indeed, recent opinion polls indicate overwhelming support for disposing of HLNW within Sweden rather than exporting them abroad. There even appears to be some public support for siting a repository in one's own region, despite the fact that Sweden has reversed its previous plan to phase out nuclear power plants. (It had been speculated that one of the reasons Sweden was able to make such progress with its disposal program was because its citizens believed the planned phase-out of nuclear plants would effectively limit the accumulation of wastes. The more plants that are built, the greater the amount of waste that will eventually require disposal.)

Finally, Sweden has maintained great flexibility in its management approach. The existence of an interim storage facility for spent fuel rods allows time to conduct characterization studies for a permanent disposal site and engage community dialogue on the project. In fact, a full decade has been allotted for discussions at prospective repository host sites to resolve issues of concern. Repository plans also stress deep disposal and multiple, long-lived, engineered barriers. This long-term role for engineered barriers (and particularly the copper canister) is consistent with the Swedish approach to err on the side of safety and to minimize long-term uncertainties. Their program contrasts with the United States approach of depending far more on the geology of the repository site than on long-lived engineered barriers to isolate wastes from the environment.

Canada

Like France and the United Kingdom, Canada has redefined and restructured its approach to nuclear waste management. Initially, Canada vested responsibility in an established federal nuclear institution, Atomic Energy of Canada Limited, and pursued a narrow technical approach in developing a management system. But strong public concern and opposition in 1977 and 1978 forced major changes to the program. The siting process was postponed so that an effort could be made to reach a consensus (nationwide and in the province of Ontario, which contains most of Canada's nuclear reactors) on an acceptable waste disposal concept. Canada's goal is to achieve this political and technical consensus by the mid- 1990s. The Canadian program had adopted an enlightened underlying philosophy stipulating that no effort to locate a site will be made until, or unless, the public accepts the disposal concept and that no attempt will be made to force a site on a community. About 3% of the total Canadian waste disposal budget has been earmarked for consultation and negotiation with the public and potential host communities (NWTRB, 1991). In addition, social science research has been commissioned to gain an understanding of the nature of public concerns. This work has highlighted such issues as public skepticism about scientific risk assessment and computer model-based predictions, ethical concerns, potential impacts on aboriginal peoples, and the need to demonstrate technology (NWTRB, 1991).

Direct federal–provincial bargaining will shape the future of the disposal program. In contrast with the United States, no legislative provision exists for breaking a federal–provincial deadlock over site selection; a resolution of differences must come from negotiation. There have also been significant institutional changes, notably the fact that Environment Canada has been given the final say as to the acceptability of the disposal concept.

Meanwhile, Canada is in no hurry to put wastes in the ground. Disposal is not expected to occur before 2025, and Canadian utilities appear prepared to provide long-term interim storage at nuclear reactor sites, although a centralized monitored facility is also a possibility. Canada's licensing and regulatory situation is also more flexible than that in the United States or Germany. Given the fact that nearly all nuclear energy in the country is produced in one province (Ontario) and by one utility (Ontario Hydro), Canada faces less management and political complexity than the United States; this, plus the restructuring that has already occurred in response to lessons learned, improves the outlook for progress in the coming decade (Sheng, Ladanyi, & Shemilt, 1993).

Generic Lessons

The European and Canadian experiences with HLNW disposal invite comparison with that of the United States (also see McKinley & McCombie, 1994, for a recent description of the Swiss HLNW program). Not only do such comparisons highlight the fact that the United States effort differs in several important ways, but comparisons also draw attention to the puzzling rigidity and lack of adaptability of the United States program in the face of some clear messages from international experience and structural issues.

The Rush to Burial

No country outside the United States and Germany is in a hurry to put HLNW in the ground. In Sweden, the successful siting and development of a central interim storage facility has provided long-term storage capacity for spent fuel rods, allowing for a cautious and deliberate search for a permanent disposal site, as well as time to resolve technical uncertainties and address social and political issues. The Swedes located the interim facility adjacent to an existing reactor, recognizing that communities experienced with hosting a nuclear plant might be more willing to accept waste storage facilities.

None of the European countries are in a hurry to find a permanent disposal site. Sweden expects to keep spent fuel rods in temporary storage for at least 40 years and will not choose a repository site until 2003–2006. France intends to spend at least 15 years studying the suitability of its test sites before making recommendations about a permanent facility. Germany and the United Kingdom also envisage lengthy interim storage of wastes before emplacement in a repository. Even Canada, which lacks a centralized interim storage facility, does not intend to recommend a site until the public has accepted the disposal concept.

Indeed, the United States is the only country in which federal law requires the national government to enter into contracts with nuclear utilities to accept waste disposal by a certain date. It is also the only country that lacks an integrated and widely-accepted plan for long-term interim storage of spent fuel rods. Many problems are associated with the loss of flexibility created by such a premature, unrealistic, schedule-driven program of waste disposal, but perhaps none are more far-reaching than the following: (1) the possible compromise of the technical integrity of the program, an issue repeatedly cited by the Nuclear Waste Technical Review Board (recently in NWTRB, 1992, 1993); (2) possible premature technical decisions; (3) choices based on expediency rather than on results of solid technical analysis, something the NWTRB (1993) says has already occurred; and (4) problems with program credibility.

Single Site Strategy

The United States and Germany are the only countries characterizing a single site for potential repository construction. This strategy leaves the waste program highly vulnerable to surprise and the risk of failure (a problem recognized early on by a program review during the Carter Administration; DOE, 1978). Recognizing this vulnerability, most countries are pursuing a strategy of multi-site consideration and characterization. Moreover, they have established research and development programs at underground facilities where work on repository-related uncertainties can proceed while the search for a site goes on. The United States lacks such a facility (NWTRB, 1992).

Engineered vs. Geologic Barriers

The United States is also the only country that does not plan to rely heavily on engineered barriers for long-term waste isolation. The use of multiple barriers—a defense-in-depth approach—that has long operated in the nuclear industry has been abandoned for the permanent HLNW repository. This abandonment greatly increases the risks associated with characterizing only a single site, because of the great reliance that will be placed on the geology of the site eventually selected. It also reduces the resilience and potential repairability of a repository.

Rigid Regulatory Environment

The United States is nearly alone in the rigidity of its highly quantified and specific regulatory approach to the waste disposal problem. The German experience does provide some parallels, however, and its regulatory framework, accompanied by a contentious process of regulatory and legal challenges, has produced a near-stalemate in the German waste management program. In contrast, most other countries allow greater agency discretion, but a discretion which, it is essential to note, rests on the philosophy of securing national political consensus and acceptance of the waste facility by the local host region.

Social Consensus and Accommodation

The nuclear weapons program was the progenitor of today's civilian nuclear energy industry. This backdrop, along with the technology's potential for catastrophic accidents, shaped the nature of the institutions, processes, and attitudes that have evolved to deal with HLNW disposal. Initially, the waste problem was conceived to be fundamentally technical in nature, and scientific research and technical assessment were adopted as the primary tools for developing solutions. The idea that the public might intrude in this process and take an active role in decision-making was anathema to many of those mandated to manage nuclear wastes. Similarly, the siting process was viewed (nowhere more strongly than in France) as a technical matter.

All this has now unraveled, and the vivid failures of this approach are apparent throughout Europe and North America. As a result, most nuclear nations are experimenting with new ways of building social acceptance. In Germany, the public can tour the site of the proposed waste repository at Gorleben and read any document relating to the project. France plans extensive negotiations with prospective host communities to resolve siting conflicts and will offer an image loss tax subsidy of millions of dollars a year to communities accepting even the underground test facilities. Substantial economic incentives, such as preferential hiring and purchasing and regional development assistance, also will be offered. Canada has committed itself to negotiation and voluntary acceptance of a repository by the host province and host region; specifically, the government has determined that it will not attempt to override a federal–provincial deadlock.

The European and Canadian experiences demonstrate that the United States is becoming increasingly isolated in its attempt to override, rather than address, state and local concerns. Most countries have determined

that success depends on developing approaches that are socially acceptable as well as technically sound, collaborative rather than preemptive,
and predicated on persuasion and negotiation rather than on coercion. In
contrast, the United States seems intent on resisting this trend toward a
more socially-acceptable method of siting a repository, as indicated by the
long delay in appointing a nuclear waste negotiator and a premature
declaration by former Energy Secretary James Watkins that the negotiating approach has failed.

Adaptation

In most countries grappling with the issue HLNW disposal issue, an
awareness has grown that the task is far different and more complex than
originally conceived. Increasing acceptance of the fact that questions of
ethics and social values must be factored into the equation, along with
strictly technical and scientific issues, has also developed. Given deep-
rooted and widespread public concerns about things nuclear, it is crucial
to gain the acceptance of prospective host communities and stakeholders
through consultation and negotiation. These mechanisms may help re-
establish public trust in the institutions managing disposal programs.

Different countries have responded in markedly different ways to these
new realities. Sweden is the clearest model of success, having developed
a strategy and created institutions that embrace flexibility, the paramount
role of social issues, a safety-first philosophy, and a process of accommo-
dation, negotiation, and consensus building. Canada has abandoned the
purely technical approach to waste management and has adopted a
model similar in many ways with the key elements of the Swedish approach. While France pursued a hardball strategy almost to the bitter end,
it is now seeking a more open and accommodative process—although it
remains to be seen how successful this process will be.

The record in other countries is even less encouraging. The United
Kingdom has essentially abrogated its responsibility to manage wastes,
handing the task off to future generations. The United States and Germany are striking in the extent to which they have recognized but failed
to respond to—much less adapt to—the hard lessons of recent years. Not
surprisingly, both programs are in serious trouble.

Ultimately, each country must find its own solutions, consistent with
the demands of its history and culture and the social needs and values
of its people. Still, important—indeed, inescapable—generic lessons can
be learned from experiences in different countries during the past decade:

- social issues must be put on an equal footing with technical matters;
- coercive approaches to siting a repository do not work; they carry

an extraordinarily high political price and provoke increasing public opposition;

- public concerns are deep-seated and unlikely to be lessened by purely technical assessments and reassurances, or by high-handed recourse to expert opinion;
- a highly integrated and adaptive management structure is both preferred and necessary;
- the entire process of siting and building a HLNW repository is likely to require a long time horizon and continuing mid-course corrections.

7

Where Do We Go from Here?

The HLNW management program in the United States is failing badly, beset by technical difficulties, poor management, scientific uncertainties, cost overruns, equivocal political support, state opposition, and profound public mistrust and antipathy. It is doubtful the existing program or its managers can overcome these obstacles in its current mission to site a HLNW repository. In particular, the unfairness of the procedures used to limit site characterization efforts to Yucca Mountain, Nevada, and the stubborn disregard for local objections to the project, virtually guarantee a continuation of the fractious, messy conflicts that have dogged the waste program from its earliest days.

It's clear that a new approach to managing the program and finding a site for the repository is urgently needed. The following recommendations outline some crucial elements of such an approach.

Reevaluate the Commitment to Underground
Geologic Disposal

Congress should place a moratorium on the current program and remove the deadline of 2010 for beginning operation of the repository. Flexible and realistic timetables would allow more time for further research to be done on technical problems associated with the repository and on comparative advantages and disadvantages of different geological structures. More effort should also be devoted to developing multiple engineered barriers to isolate HLNW from the environment. A moratorium on geologic disposal would also allow time to more carefully evaluate alternative techniques, such as seabed disposal.

More importantly, delaying the disposal program would create an opportunity for the federal government to make a genuine effort to gain public acceptance and political support for the program. Delay would provide the leeway needed to establish a voluntary process for selecting a repository site.

Use Interim Storage Facilities

These facilities would buy time. Above ground storage in dry casks at reactor sites or a centralized monitored retrievable storage (MRS) facility could be used to store wastes for 100 years or more. Although not without problems, MRS would cost considerably less than a permanent geologic repository and would buy time for additional research and public consultation processes described above. The federal government should therefore abandon its current policy prohibiting the development of an MRS facility until a site for a permanent underground repository is found.

Evaluate More than One Site

By evaluating more than one site program failure will be less likely. Every effort must be made to find several states and communities willing to be considered as the location of an interim or permanent storage facility. It is crucial to keep several options open until very late in the selection process, because the repository is a first-of-its-kind facility with a great many associated uncertainties and a well-demonstrated ability to evoke intense public and political opposition. The arbitrary selection of Yucca Mountain as the only site to be characterized closed off all other options, which, given Nevada's strong and unrelenting resistance to the project, creates the very real possibility that the nation may be left without any likely repository site.

Characterizing a single site also increases risks associated with relying most heavily on geologic structures, rather than engineered barriers to isolate wastes. The former is currently the approach taken in the United States program. If the geologic barrier at the chosen site does not live up to expectations, a long and laborious site selection process would again have to be geared up from square one.

Employ a Voluntary Site Selection Process

The arbitrary and unfair procedures used to select Yucca Mountain for site characterization have been a major source of conflict and have evoked fierce public and political opposition to the project. To avoid such conflicts in the future, Congress should mandate that no community will be forced to accept a repository against its will and that potential host communities should be encouraged and permitted to play a genuine and active role in the planning, design, and evaluation of the repository. Experience from other countries shows that it is possible to develop an effective siting process that encourages such local participation. If more than one community is willing to participate in a voluntary program, it is even possible that a competitive process could result.

The approach currently taken by the United States nuclear waste negotiator to site an MRS facility is sensible; interested communities are given planning grants that do not carry with them an obligation to host the facility. These grants enable communities to learn about the technical aspects of the storage process and to determine whether local residents are really interested in hosting the facility.

A voluntary process requires not only public participation, but also an agreed-on procedure (e.g., public referendum with a two-thirds plurality) for determining whether to accept the facility. Such a voluntary process must be given every chance to work. If no community or region is willing to accept a HLNW facility, some other siting method must be found, but it must be based on principles of openness and fairness.

Negotiate Agreements and Compensation Packages

A voluntary MRS or repository siting program must offer sufficient benefits to potential host communities and regions so that their residents feel their situation has improved over the status quo. This compensation has not occurred in the case of Yucca Mountain. The federal government has offered Nevada $10 million a year during the characterization phase and $20 million a year during the operation of the repository, but nearly 66% of Nevada residents questioned in a 1993 survey believed that the harms of the repository would outweigh the benefits it would provide.

Acknowledge and Accept the Legitimacy of Public Concerns

The attitude that HLNW disposal is merely a technical problem to be solved by experts must be abandoned. A repository program has social and economic dimensions that will seriously affect the quality of life in neighboring communities. Most notably, such a project has the potential to stigmatize these communities, making them less attractive to residents, visitors, businesses, and in-migrants. Negotiated compensation packages should take into account the fact that stigma effects could have extremely negative long-term economic and social consequences in the affected communities.

Guarantee Stringent Safety Standards

Many people associate nuclear wastes with danger and death and react to the idea of a HLNW repository with feelings of fear and dread. Their trust in the ability of waste managers to protect them from danger is not enhanced by political moves like the 1992 Energy Policy Act, which set less stringent radiation-exposure standards for Yucca Mountain than had originally been contemplated.

Public acceptance of the repository program requires assurances that public safety will be a priority. The federal government must negotiate contingent agreements with any community or region that agrees to host a repository and specify what actions will be taken should there be accidents or unforeseen events, interruptions of service, changes in standards, or the emergence of new scientific information about risks or impacts.

The use of interim storage technologies, as recommended earlier, offers some safety advantages over a permanent repository. With both reactor-site storage and a central MRS facility, wastes are kept above ground; it should be easier to deal with any problems that occur because the wastes are not buried deep underground. On the other hand, if anything happens to the repository's underground containers after the facility is permanently sealed, it would be difficult and expensive to reach them.

Restore Credibility to the Waste Disposal Program

By creating a new management organization and adopting a new management approach credibility may be restored. The history of the national HLNW management program to date underscores one glaringly obvious point: DOE has failed in its management role and is incapable of overseeing such an extraordinarily complex and uncertain program. Nor can DOE overcome widespread mistrust and skepticism about its competence, methods, and motives. Consequently, Congress should establish a new agency or organization to manage the civilian HLNW program, which should be separated from the military program. Not only have DOE's problems with military wastes severely compromised its credibility, but the job of cleaning up contaminated defense sites is vastly different than siting and managing a HLNW facility for civilian reactor wastes.

No matter which agency is assigned the job, however, a radical new management approach is needed—one committed to implementing the recommendations listed above and doing so in an open, consultative, and cooperative manner that does not seek to deny or avoid the serious social and economic dimensions of the HLNW disposal problem.

Given large inherent uncertainties with the repository program and the likelihood that it will encounter unforeseen surprises, the management approach must emphasize flexibility and adaptability. Only by frankly acknowledging the limitations of the available information and predictive techniques can program managers promote social trust and elicit a readiness to tackle the problem from a different direction if necessary.

This strategy requires a cautious and deliberate approach. Repository development must proceed in stages, allowing time to discover whether the predictions and expectations associated with the project are reasonable. Assessment techniques and models must be viewed as learning tools, not as crystal balls that can foretell the future with precision. And because we have no analytical tools that can completely eliminate the element of surprise, we must develop methods for coping with it. Such methods include (1) monitoring systems that tell us what is actually happening and help us learn as we go; (2) redundancy in systems designed to protect; (3) technical reversibility or repairability; (4) multiple geologic and engineered barriers; and finally, (5) multiple disposal sites.

Conducting the program in a careful, step-by-step fashion will permit—in fact, encourage—everyone to stop and reflect and, if necessary, make mid-course corrections mandated by new technical developments and changing social values. All of these measures will enhance the adaptability of the management system, reduce the possible consequences of any mistakes and mishaps that occur along the way, and increase the chances of recovering—and learning—from mistakes while retaining public support and confidence.

In addition to adaptability, the new management approach requires reliability (do the job without error), durability (survive for the length of time needed to complete the job), stability (continue performing the required tasks despite external changes), capacity (handle the required volume of wastes) and integrative skills (consider system-wide consequences of decisions).

Above all, we must remember that building a permanent, underground HLNW repository is essentially an experiment—one whose full social and economic dimensions are uncertain and unpredictable and likely to remain so. Citizens must decide how—and perhaps even whether—to proceed with this experiment in the face of all the unknowns and potential risks it presents.

Only by conducting the experiment in a conspicuously fair manner and making every effort to achieve social consensus about what to do can the federal government hope to rise above the mistakes and miscalculations of the past.

8

Update on the High-Level Nuclear Waste Repository Siting

Publishing a book takes time. We are fortunate that with Westview Press the total time from submission of the manuscript to the finished book was less than a year. However, this time lag raises the question of how to deal with the most recent events and changes in the United State's repository program. This chapter updates the book with discussion of some important developments concerning the high-level nuclear waste (HLNW) program.

Problems and Recommendations

One area with continual problems is the intergovernmental relations involving DOE and state and local governments. The U.S. Secretary of Energy Advisory Board Task Force on Radioactive Waste Management released a final draft report (November 1993), *Earning Public Trust and Confidence*, with a number of useful recommendations, including the recommendation that states and other stakeholders be substantially involved in DOE programs. These recommendations have not yet been applied to DOE relations with the State of Nevada and the affected local governments of the Yucca Mountain project.

In late 1993 DOE's Office of Civilian Radioactive Waste Management (OCRWM) released a new draft public involvement policy that called for substantive two-way communication with stakeholders (DOE, 1993). This document was similar to other official DOE statements put forward since 1985. As a State of Nevada review noted, to be meaningful communication must "provide the opportunity for the public to influence decisions" (NWPO, 1994, p. 3). As if to underscore the aptness of these state comments and the difficulties facing enactment of its communication goals, DOE announced the Proposed Program Approach (PPA), a major change in program design, to a nuclear industry group in May 1994 without any prior consultation or communication with the State of Nevada or any nonindustry stakeholders. (This change in program approach is discussed further on pages 106–108.)

DOE was successful in obtaining a 40% increase in their budget for fiscal year 1995. However, the allocation to the affected local governments and the State of Nevada were not increased. The allocation to the State of Nevada is significantly less than it was in 1988, even though the state and the affected local governments are the only nonfederal government entities that have legal responsibility for oversight and assessment of the Yucca Mountain project.

Two legal actions by the State of Nevada should be noted. A 1994 Nevada suit, *Nevada v. O'Leary* (94–70148: 9th Circular Court), claimed that DOE failed to adequately characterize the nature and origin of calcite-silica deposits in Trench 14 at Yucca Mountain. The import of this case is that the Nuclear Waste Policy Act (NWPA) requires DOE to terminate all site characterization activities if at any time Yucca Mountain is found to be unsuitable. DOE, however, does not have procedures for making a finding of unsuitability as a result of their site characterization studies. Nevada argues that the decision by DOE to discontinue study of the possible disqualifying conditions put forward in the Szymanski model, a strongly contested technical issue between the state and DOE, is a failure to carry out appropriate site characterization activities at Yucca Mountain. The Szymanski theory postulates that geologic and hydrologic processes (coupled processes) have in the past caused groundwater to rise within Yucca Mountain causing significant calcite-silica deposits. This model goes on to state that these interactive processes could cause in the future a similar rise in groundwater within the repository space. If this rise were to happen and groundwater came into contact with the extremely hot and radioactive waste containers, the resulting high pressure steam and/or direct radioactive contamination of the water source with the HLNW could result in a catastrophic accident. The state asks the court to establish the state's right to a credible scientific process.

Another suit now before the 9th Circuit Court also addresses the issues of how DOE decisions are made and what record will be maintained. This case, *Nevada v. O'Leary*, asks that the state be allowed to take depositions from 27 scientists who reviewed the studies of the Szymanski theory.

An important area for DOE has been with the management of the HLNW program and the Yucca Mountain project. Many complaints from a variety of critics have been about the cost of the program and the slow pace of progress. A recent example was the failure of project managers to order the conveyor equipment to match the capacity of the new tunnel boring machine to be used in excavation of the underground exploratory studies facility. The result is that tunnel work will go forward at only about one-third the rate possible with the proper equipment, and ordering and installing a proper conveyor system is expected to take more than a year. The overall result of past examples of management problems has been a series of recommendations and calls for substantial changes at DOE.

These recommendations seem to increase as it becomes clear that the last legislative or reorganizational "fix" has failed to put the program back on track. During 1993 and 1994 there were many calls for new examination of the DOE program. A November 1993 letter to the U.S. Secretary of Energy from the Nuclear Waste Technical Review Board, which was established by Congress to review the scientific and technical aspects of the repository program, suggested a major program review. This letter was followed by an official recommendation in February 1994. In March 1994 ten senators asked President Clinton to appoint a special commission, independent of DOE, to review the nation's nuclear waste programs. These senators pointed to budget, management, schedule, and public acceptance problems with low-level radioactive wastes, temporary waste storage facilities, handling of transuranic wastes, and the HLNW repository program. They complained that the current federal efforts are piecemeal, not cost-effective, and unable to provide acceptable nuclear waste management at any level. In 1993 at least a dozen calls were made for investigation and review of the Yucca Mountain program from a wide range of individuals and groups (see NWPO, n.d.). Nevada congressional representatives asked for a review in June 1993, followed by a similar request by the Nevada governor in July. The Western Governors' Association request for a review was contained in a June 1993 resolution. Several public interest groups joined in July 1993 and asked President Clinton to initiate a comprehensive review. Congressional members sent letters to the U.S. Secretary of Energy and to the president in August and October 1993. In August 1994, the Nuclear Waste Strategy Coalition, consisting of state regulators, utility executives, and attorney general representatives, released a draft report, *Redesigning the U.S. High-Level Nuclear Waste Disposal Program for Effective Management* (1994). A coalition working group reviewed 18 federal government documents produced between 1982 and 1994 on the subject of DOE management issues and problems. The report recommended that DOE be removed from HLNW management and their duties be reassigned to a federally-chartered corporation.

In 1993 the U.S. Secretary of Energy commissioned "an independent review of the financial and management" performance of DOE's work on the Yucca Mountain project. DOE appointed one person to oversee the evaluation, a private industry executive; the Nevada governor appointed the other person, a Nevada Public Service Commissioner. These overseers asked the General Accounting Office (GAO) to comment on the statement of work prepared by DOE. This statement would serve as the basis for a contract to a consulting firm who would conduct the study. The GAO report (1994a) said that the review proposed by DOE "is too narrow and may result in a product that, while useful, will not address many of the major issues confronting the disposal program" (p. 3). GAO also noted that the funding and time allowed for the review work were probably inadequate. At this point it is not clear when, or whether, this review will be completed.

GAO reported in May 1993 that more that 60% of allocated funds were being spent on infrastructure activities (i.e., management and administration) and only about 22% on essential scientific and technical activities at Yucca Mountain. GAO recommended a thorough review of the program. In a September 1994 report GAO said that recent DOE review initiatives "are too narrow in scope and lack sufficient objectivity to provide the thoughtful and thorough evaluation of the program that is needed" (p. 1). The report states that "an independent review is needed now more than ever" (p. 8), and it claims that the "issues that need to be resolved are fundamental and need to be addressed objectively" (p. 11). GAO recalls that it recommended a comprehensive review as far back as 1991. In explaining some of its concerns, GAO commented on the new DOE multipurpose container (MPC) system, which combines scientific, technical, and management problems.

The Clinton administration, elected in November 1992, took office in January 1993. Over the following months they filled the key offices in DOE. Daniel Dreyfus began work at DOE in February 1993, although he was not confirmed as the director of the Office of Civilian Radioactive Waste Management (i.e., the Yucca Mountain project) until October 1993. In interviews with the industry publication, *Nuclear Energy,* Dreyfus recounts that the HLNW program was in disarray and that within DOE there was "widespread concern that the program could not realize the expectations of its clients," which the article identifies as "the electric utilities that operate nuclear power plants and their customers" (Special Report, 1994, pp. 18–19).

The Department of Energy's Response:
The Proposed Program Approach

In response to the calls for actions, DOE made a series of revisions in its program. These changes are based on an analysis that identified three major problems. First, DOE determined that the work at Yucca Mountain was not properly focused. Second, DOE found that it needed to improve its program management. Third, DOE had to do something substantial to meet the 1998 goal of beginning to accept HLNW from the utilities. It might be noted that this 1998 date is only one of many dates mandated by Congress in the NWPA that is still considered by DOE to be operative, all other dates having long since been exceeded or rescheduled.

This recent DOE review concluded that existing program obligations were beyond the department's abilities and funding resources. In short, DOE could no longer continue as it had planned and still meet the 1998 date to begin accepting HLNW from the utilities or the 2010 date to begin operation of a repository at Yucca Mountain. This reality about the program and its schedule has been pointed out by people outside DOE many times. In May 1993, GAO estimated that the repository operations were 5 to 13 years behind the scheduled 2010 opening date. Both the National Academy of Sciences and the Nuclear Waste Technical Review Board have cautioned that DOE's schedule objectives are unrealistic and are inappropriately driving the program.

In the spring of 1994, DOE proposed a sweeping restructuring of the HLNW program. The new Proposed Program Approach (PPA) seeks to provide waste acceptance by 1998 with the use of multiple purpose canisters (MPC), which DOE would design, produce, and provide to utilities with nuclear reactors. The MPC would allow HLNW to be stored on-site at the reactors, to be shipped to an interim storage facility, and, potentially, to be used for permanent storage at a repository—all in the same canister. This approach envisions the use of specially designed overpacks for storage, transport, and disposal. However, as GAO pointed out in its September 1994 report, one problem is that not enough is known about the Yucca Mountain site to design a waste canister. In the words of the GAO report, "The Nuclear Waste Technical Review Board and others have repeatedly pointed out, and DOE program managers have acknowledged, that more information about the potential repository site and the potential effects of the heat from the waste on the repository will be needed before a disposal package with a high degree of safety assurance can be developed" (p. 7). The GAO report concluded that DOE runs the risk of having to redesign the waste canister, provide a new engineered barrier systems, or "accept certain safety risks" (p. 7). Of course, it is not only DOE that might be asked to accept certain safety risks.

In addition, the PPA would make major changes in scientific and technical work on site characterization, site suitability determination, licensing under Nuclear Regulatory Commission (NRC) regulations, and repository construction and operation. Instead of an 8 to 10 year study process to demonstrate reasonable assurance (NRC's licensing standard) that Yucca Mountain can isolate HLNW for the required 10,000 years, DOE would make an early determination of technical suitability by 1998. This determination would use available data with expert judgments to compensate in areas where data are not available or sufficient. Based on a finding of technical suitability, DOE would seek a NRC license to construct and operate the facility. A final suitability determination of the site as a repository would be postponed for 50 to 100 years to a performance confirmation period. During this period, DOE would attempt to collect data needed to fully demonstrate that Yucca Mountain complies with NRC and Environmental Protection Agency (EPA) regulations for waste isolation and reasonable assurance of repository performance.

PPA is a significant departure from past DOE planning to study Yucca Mountain and obtain authorizations and licenses for construction and operation of the site as a HLNW facility. DOE maintains that PPA does not require changes to NWPA or to NRC regulations. The State of Nevada contends the opposite—that PPA requires changes in NWPA and in NRC licensing regulations. The state further contends that PPA is a veiled attempt to build the repository and emplace HLNW at Yucca Mountain without first addressing the serious performance deficiencies the state believes to be present at the site. DOE, on the other hand, presents PPA as an innovative and cost-effective alternative to the existing approach.

PPA attempts to address schedule, cost, and some issues of scientific doubt and uncertainty associated with Yucca Mountain. It doesn't confront key institutional, organizational, and management issues discussed in the preceding chapters. Issues of public and intergovernmental participation, public acceptance, fairness and equity, voluntarism, and the need for improved management of the nation's HLNW are not addressed in this latest program design. The new approach continues to rely on non-voluntary siting of a facility at Yucca Mountain by relaxing standards for scientific study, analyses of data and information, and licensing requirements. PPA does not address the need for alternatives to Yucca Mountain in the event the site proves unsuitable. At best PPA removes the final suitability determination forward for as much as a century. Converting Yucca Mountain to a *defacto* interim storage site would address the DOE perceived need for an immediate national storage site for HLNW. However, this conversion is done in a way that creates more uncertainty even while DOE avoids fully examining the existing uncertainties. What precedent would the federal government set in short-cutting the requirements for licensing HLNW facilities? And how would this influence future public acceptance for other radioactive waste facilities or even other hazardous technologies that are subject to government regulation?

Summarizing the Situation

As this book goes to press, the issues and uncertainties described in the preceding chapters continue with the most recent attempts of the federal government to fix the HLNW program. Notwithstanding changes suggested by the Proposed Program Approach, the recommendations in chapter 7 continue to provide a useful, constructive, and potentially successful framework for HLNW management policy over the long run.

Let us summarize the situation. HLNW can be stored safely for an interim period of 100 years or more using dry-cask technology. The doubts and uncertainties presented by Yucca Mountain cannot be addressed unless we downgrade our standards for studying and licensing a repository. Yet these repository standards have been developed over the past decade as essential protection for human life and the environment.

The potential for scientific surprises and controversies is almost certain in a program dominated by high-stake risks and uncertainties. On Sunday, March 5, 1995, the headlines in the *New York Times* reported that "scientists fear atomic explosion of buried waste." The subtitle noted the "debate by researchers" and provided the opinion that the "argument strikes a new blow against a proposal for a repository in Nevada" (Broad, 1995a). The source of this controversy, as in the case of the Szymanski theory, is DOE scientists. Drs. Charles Bowman and Francesco Venneri, physicists at Los Alamos, initiated this scientific debate with the theory that the geologic structure of Yucca Mountain could slow the plutonium neutrons generated in the HLNW and initiate an atomic explosion. The results of this explosion could be catastrophic. However, this theory has been hotly debated within Los Alamos and DOE. According to Broad, a review team of 30 Los Alamo scientists, along with a few scientists outside Los Alamos, say "the explosion thesis is provocative and probably wrong," because they found the theory to be without technical merit. Bowman and Venneri, however, argue that the review resulted in a more convincing formulation of the theory because they then addressed questions and objections raised during the review process.

Less than 3 weeks later, the explosion theory was back in the news. An internal report by three senior scientists at the DOE Savannah River Site near Aiken, South Carolina, strongly endorsed Bowman and Venneri's position. They added that under certain conditions 50 metric tons of the waste plutonium could produce "an explosion equal to that of a large hydrogen bomb" (Broad, 1995b).

How much effect this controversy has on the Yucca Mountain project remains to be seen, but this issue is only one of a plentiful supply of scientific controversies that fuel a growing discontent with the assertion that Yucca Mountain is the nation's HLNW disposal solution. Several new legislative initiatives in Congress would downgrade the priority of the Yucca Mountain project and upgrade an interim storage facility, which would bring public attention to issues of storage, cask design, and transportation. Such major adjustments in the nation's HLNW program may change the terms of the current impasse between the federal program and the public, community, and state positions. However, unless the basic underlying issues and concerns are addressed in a cooperative and publicly acceptable way, these new efforts will only produce additional examples of conflict and potential failure.

The rush to build at Yucca Mountain compromises our future ability to achieve acceptable disposal of HLNW. Yucca Mountain very well could be unacceptable on any terms. There are no alternative plans, only the directions in the Nuclear Waste Policy Act to return to Congress for further instructions if Yucca Mountain is unsuitable. If Yucca Mountain is found to be unsuitable in 30, 40, 50, or 100 years, because no genuine site study was conducted, what options will exist? The nuclear power plants that produce the money for the HLNW program will have long been closed, and they no longer will be a source of funding. The Nuclear Waste Fund will have been spent. Loading Yucca Mountain with HLNW as if it were a repository, especially on compromised standards, and then finding that Yucca Mountain is unsuitable or that some other option for HLNW management is necessary, will place tremendous burdens on future generations, complicating rather than simplifying their abilities to manage the legacy of HLNW.

The fact that Yucca Mountain is a failed program must be recognized. This message will prompt resistance from DOE, the nuclear industry, and their supporters because it substantiates outside claims that the 1987 decision to look only at Yucca Mountain was a mistake.

On a positive note, the past decade has taught us many things about managing nuclear wastes. It is possible to store HLNW safely for a century or more with dry-cask technology. It is clear that there must be public support and acceptance for HLNW facilities, not just theoretically at the congressional level but at the community, state, and public levels as well. Much more is understood about public concerns. We have begun to learn how important it is to have cooperation and agreement among communities, states, and the federal government. We know that we must have an institution in charge of the HLNW management that is trusted and competent. It is time to face the fact that we must have a new national policy on HLNW. There is time and ample reason to rethink this program and come to a better solution than Yucca Mountain with all the risks and vulnerabilities it presents. This is important. It is worth doing right.

Glossary

ANEC: American Nuclear Energy Council

DOE: United States Department of Energy

EPA: United States Environmental Projection Agency

GAO: United States General Accounting Office

HLNW: High-level nuclear waste

INEL: Idaho National Engineering Laboratory

KBS: Kärnbränslesäkerhet—Swedish Nuclear Waste Management Project

MPC: Multipurpose container

MRS: Monitored retrievable storage

NIREX: Nuclear Industry Radioactive Waste Executive

NRC: United States Nuclear Regulatory Commission

NRC/NAS: National Research Council/National Academy of Sciences

NNWC: Nevada Nuclear Waste Commission

NWPA: Nuclear Waste Policy Act of 1982

NWPAA: Nuclear Waste Policy Act Amendments of 1987

NWPO: State of Nevada Agency for Nuclear Projects/Nevada Nuclear Waste Project Office

NWTRB: Nuclear Waste Technical Review Board

OTA: United States Office of Technological Assessment

PPA: Proposed Program Approach suggested by DOE

SKB: Svensk Kärnbränslehantering AB—Swedish Nuclear Fuel and Waste Management Company

SKN: Statens Kärnbränslenamnd—Swedish National Board for Spent Nuclear Fuel

WIPP: Waste Isolation Pilot Project near Carlsbad, New Mexico

Bibliography

Ahearne, J.F. (1990, October 5). *Nuclear waste disposal: Can there be a resolution?: Past problems and future solutions.* Paper presented at MIT International Conference on the Next Generation of Nuclear Power.

Ahlström, P.-E. (1994, Fall). Sweden takes steps to solve waste problem. *Forum for Applied Research and Public Policy,* 119–121.

Blowers, A., Lowry, D., & Solomon, B.D. (1991). *The international politics of nuclear waste.* New York: St. Martin's.

Broad, W.J (1990, November 18). A mountain of trouble. *New York Times Magazine,* pp. 37–39, 80–82.

Broad, W.J. (1995a, March 5). Scientists fear atomic explosion of buried wastes. *New York Times,* p. 1.

Broad, W.J. (1995b, March 23). Theory on blast threat at nuclear dump gains support. *New York Times,* p. 1.

Carter, L.C. (1987). *Nuclear imperatives and public trust: Dealing with radioactive waste.* Washington, DC: Resources for the Future.

Carter, L.J. (1993, Fall). Ending the gridlock on nuclear waste storage. *Issues in Science and Technology,* pp. 73–79.

City of Sante Fe, New Mexico, v. Komis. (1992, August 26). Appeal from the district court of Sante Fe County (Case No. 20,325).

Colglazier, E.W. (1982). *The politics of nuclear waste.* Elmsford, NY: Pergamon.

————, & Langum, R.B. (1988). Policy conflicts in the process of siting nuclear waste repositories. *Annual Review of Energy,* **13,** 317–357.

Cook, B.J., Emel, J.L., & Kasperson, R.E. (1990). Organizing and managing radioactive waste disposal as an experiment. *Journal of Policy Analysis and Management,* 9(Summer), 339–366.

————. (1991/92). A problem of politics or technique? Insights from waste-management strategies in Sweden and France. *Policy Studies Review,* **Winter,** 103–113.

————. (1992). *Clashing judgments, common fears.* Unpublished monograph, Clark University, Worcester, MA.

Dantico, M., & Mushkatel, A. (1991). Governors and nuclear waste: Showdown in the rockies. In E.B. Herzik & B.W. Brown (Eds.), *Gubernatorial leadership and state policy.* New York: Greenwood.

Davis, J.A. (1988). The wasting of Nevada. *Sierra,* **73,** 30–35.

Desvousges, W.H., Kunreuther, H., Slovic, P., & Rosa, E.A. (1993). Perceived risk and attitudes toward nuclear wastes: National and Nevada perspectives. In R. Dunlap, M.E. Kraft, & E.A. Rosa (Eds.), *Public reactions to nuclear waste: Citi-*

zens' views of repository siting (pp. 175–208). Durham, NC: Duke University Press.

Dingell, J. (1993, February 17). Opening statement of Chairman of the Subcommitte on Oversight and Investigations of the House Committee on Energy and Commerce. U.S. Congress: GAO.

Easterling, D., & Kunreuther, H. (1995). *The dilemma of siting a high-level radioactive waste repository.* Boston: Kluwer.

Edison Electric Institute. (1992, November). *Report on the eighth review of the Yucca Mountain project, U.S. Department of Energy.* Washington, DC: Author.

Energy Policy Act. Pub. L. No. 102–486 (1992, October 24).

Erikson, K. (1990). Toxic reckoning: Business faces a new kind of fear. *Harvard Business Review,* **90**(1), 118–126.

———. (1994, March 6). Out of sight, out of our minds. *New York Times Magazine,* pp. 34–41, 50, 63.

———, Colglazier, E.W., & White, G.F. (1994, Fall). Nuclear waste's human dimension. *Forum for Applied Research and Public Policy,* pp. 91–97.

Fiske, S.T., Pratto, F., & Pavelchak, M.A. (1983). Citizen's images of nuclear war: Contents and consequences. *Journal of Social Issues,* **39,** 41–65.

Flynn, J. (1992). How not to sell a nuclear waste dump. *Wall Street Journal,* p. A20.

———, Kasperson, R., Kunreuther, H., & Slovic, P. (1992). Time to rethink nuclear waste storage. *Issues in Science and Technology,* **8**(4), 42–48.

———, Slovic, P., & Mertz, C.K. (1993). The Nevada initiative: A risk communication fiasco. *Risk Analysis,* **13,** 643–648.

Frieden, B.J., & Kaplan, M. (1975). *The politics of neglect: Urban aid from model cities to revenue sharing.* Cambridge, MA: MIT University Press.

Galpin, F.L., & Clark, R.L. (1991). EPA's development of environmental standards for high-level and transuranic waste. In *High-level radioactive waste management: Proceedings of the second annual international conference* (pp. 1–6). La Grange Park, IL: American Nuclear Society and American Society of Civil Engineers.

German, J. (1993, July 2). Senate nukes Yucca resolution. *Las Vegas Sun.*

Graham, B. (1992, October 8). Cong. Rec. S. §17569 *et seq.*

Greenwood, M., McClelland, G., & Schulze, W. (1988). *The effects of perceptions of hazardous waste on migration* (NWPO–RP–0083). Carson City, NV: NWPO.

Hansson, S.O. (1987). *Risk decisions and nuclear waste* (SKN report 19). Stockholm: Swedish National Board for Spent Fuel.

———. (1992). Decision-making under great uncertainty. In *Nuclear waste management review work—Part of the decision making process. Proceedings from a symposium* (pp. 99–110; SKN report 55). Stockholm: Swedish National Board for Spent Fuel.

Hardin, E., & Ahagen, H. (1992). *Survey of siting practices for selected management projects in seven countries* (SKN report 60). Stockholm: Swedish National Board for Spent Nuclear Fuel.

Healy, M. (1993, December 8). U.S. reveals secret nuke experiments. *Register-Guard,* p. A1.

Herzik, E.B., & Mushkatel, A.H. (1988). *Urban area intergovernmental studies report* (MRDB: IG0006). Carson City, NV: NWPO.

———. (1991/92). Intergovernmental complexity in nuclear waste disposal policy: The indeterminate role of local government. *Policy Studies Review,* 10(4), 139–151.

Holdren, J.P. (1992). Radioactive-waste management in the United States: Evolving policy prospects and dilemmas. *Annual Review of Energy and the Environment,* 17, 235–259.

Holmberg, S. (1991). *The impact of party on nuclear power attitudes in Sweden* (SKN report 48). Stockholm: Swedish National Board for Spent Fuel.

Jacob, G. (1990). *Site unseen: The politics of siting a nuclear waste repository.* Pittsburgh, PA: University of Pittsburgh Press.

Judd, D.R. (1988). *The politics of American cities: Private power and public policy* (3rd ed.). Boston: Scott, Foresman, & Co.

Kasperson, R.E. (1992). The social amplification of risk: Progress in developing an integrative framework. In S. Krimsky & D. Golding (Eds.), *Social theories of risk* (pp. 153–178). Westport, CN: Praeger.

_____, & Kasperson, J.X. (1987). *Nuclear risk analysis in comparative perspective: The impacts of large-scale risk assessment in five countries.* Boston: Allen & Unwin.

_____, Ratick, S., & Renn, O. (1988). *A framework for analyzing and responding to the equity problems involved in high-level radioactive waste disposal* (NWPO–SE–019–89). Carson City, NV: Nevada Nuclear Waste Project Office.

Keeney, R.L., & von Winterfeldt, D. (1994). Managing nuclear waste from power plants. *Risk Analysis,* 14(1), 107–130.

Kraft, M. (1992). Public and state responses to high-level nuclear waste disposal: Learning from policy failure. *Policy Studies Review,* 10(4), 152–166.

Kunreuther, H., Desvousges, W., & Slovic, P. (1988). Nevada's predicament: Public perceptions of risk from the proposed nuclear waste repository. *Environment,* 30(8), 16–20, 30–33.

_____, & Easterling, D. (1992). Gaining acceptance for noxious facilities with economic incentives. In D.W. Bromley & K. Segerson (Eds.), *The social response to environmental risk: Policy formulation in an age of uncertainty.* Boston: Kluwer.

_____, Easterling, D., Desvousges, W., & Slovic, P. (1990). Public attitudes toward siting a high-level nuclear waste repository in Nevada. *Risk Analysis,* 10, 469–484.

_____, Fitzgerald, K., & Aarts, T. (1993). Siting noxious facilities: A test of the facility siting credo. *Risk Analysis,* 13(3), 301–318.

Lenssen, N. (1991, December). *Nuclear waste: The problem that won't go away* (Worldwatch paper 106). Washington, DC: Worldwatch Institute.

McKay, B., & Swainston, H.W. (1989, November 1). Letter to Nevada Governor Robert Miller (21 pp.). Carson City, NV: Author.

McKinley, I., & McCombie, C. (1994, Fall). Switzerland plans to bury nuclear waste problem. *Forum for Applied Research and Public Policy,* pp. 116–118.

Merkhofer, M.W., & Keeney, R.L. (1987). A multiattribute utility analysis of alternative sites for the disposal of nuclear waste. *Risk Analysis,* 7, 173–194.

Mitchell, T., & Scott, W. (1987). Leadership failures, the distrusting public, and prospects of the administrative state. *Public Administration Review,* 47, 445–452.

Mountain West. (1989, June). *Yucca Mountain socioeconomic project: An interim report* (NWPO–SE–024–89). Carson City, NV: NWPO.

Mushkatel, A.H. (1987). *Intergovernmental relations and local cost analysis.* Carson

City, NV: NWPO.

————, & Pijawka, K.D. (1992). *Institutional trust, information, and risk perceptions* (NWPO–SE–055–92). Carson City, NV: NWPO.

National Research Council/National Academy of Sciences Board on Radioactive Waste Management. (1990). *Rethinking high-level radioactive waste disposal: A position statement of the board on radioactive waste management.* Washington, DC: National Academy Press.

————. (1992, April). *Ground water at Yucca Mountain: How high can it rise?* Washington, DC: National Academy Press.

Nelkin, D. (Ed.) (1992). *Controversy: Politics of technical decisions.* Newbury Park, CA: Sage.

Nevada Nuclear Waste Project Office (n.d.) *Collection of calls for an independent, comprehensive review of the Department of Energy's civilian radioactive waste program* (reprints of reports, testimony, letters, etc.). Carson City, NV: Author.

————. (1994, January 13). *State of Nevada comments on the U.S. Department of Energy's draft public involvement policy.* Carson City, NV: Author.

Nevada v. O'Leary, 151 F.R.D. 655 D. Nev. (1993).

Nevada v. O'Leary, No. 94–70148, 9th Circuit Court of Appeals. (1994).

Nevada v. Watkins, 914 F.2d.1545. (1990).

Nuclear Waste Policy Act Amendments of 1987, Title V—Energy and environment programs, Subtitle A—Nuclear waste amendments. Pub. L. No. 100–203, 101 Stat. 1330–227. (1987).

Nuclear Waste Policy Act of 1982, 42 U.S.C. § 10101–10226, Pub. L. No. 97–425.

Nuclear Waste Strategy Coalition. (1994, August 28). *Redesigning the U.S. high-level nuclear waste disposal program for effective management* (draft). St. Paul, MN: Minnesota Department of Public Service.

Nuclear Waste Technical Review Board. (1991, December). *Fourth report to the U.S. Congress and the U.S. Secretary of Energy* (0–16–036006–4). Washington, DC: GPO.

————. (1992, December). *Sixth report to the U.S. Congress and the U.S. Secretary of Energy.* Washington, DC: GPO.

————. (1993, March). *NWTRB special report to Congress and the Secretary of Energy.* Arlington, VA: Author.

Omnibus Budget Reconciliation Act of 1987. Pub. L. No. 100–202, 101 Stat. 1329 (1987) and Pub. L. No. 100–203, 101 Stat. 1330 (1987).

Openshaw, S., Carver, S., & Fernie, J. (1989). *Britain's nuclear waste: Siting and safety.* London: Bellhaven.

Pijawka, K.D., & Mushkatel, A.H. (1992). Public opposition to the siting of the high-level nuclear waste repository: The importance of trust. *Policy Studies Review,* **10**(4), 180–194.

Rhodes, J. (1990, November 13). *Nuclear power: Waste disposal: New reactor technology, pyramids underground.* Paper presented at the 102nd Annual Meeting of the National Association of Regulatory Utility Commissioners, Orlando, FL.

Rogers, K. (1992, June 30). Quake rattles nuke dump's future. *Las Vegas Review-Journal,* p. A1.

————. (1993, July 3). GAO probe set on nuke dump. *Las Vegas Review-Journal,* pp. 1A, 4A, 5A.

Rydell, N. (1992). Swedish experience in spent fuel research—A reviewer's perspective. In *Nuclear waste management review work—Part of the decision making process. Proceedings of a symposium* (pp. 17–22; SKN report 55). Stockholm: Swedish Board for Spent Nuclear Fuel.

Salpukas, A. (1993, April 13). U.S. seeks help in finding site for nuclear waste. *New York Times*, pp. A9.

Sheng, G., Ladanyi, B., & Shemilt, L. (1993). Canada's nuclear waste concept—The evaluation process. *Energy Studies Review, 5*(3), 165–179.

Shrader-Frechette, K.S. (1993). *Burying uncertainty: Risk and the case against geological disposal of nuclear waste.* Berkeley: University of California Press.

Shulman, S. (1992). *The threat at home: Confronting the toxic legacy of the U.S. military.* Boston: Beacon.

Slovic, P., Layman, M., Kraus, N., Flynn, J., Chalmers, J., & Gesell, G. (1991). Perceived risk, stigma, and potential economic impacts of a high-level nuclear waste repository in Nevada. *Risk Analysis, 11*(4), 683–696.

_____ , Lichtenstein, S., & Fischhoff, B. (1979). Images of disaster: Perception and acceptance of risks from nuclear power. In G. Goodman & W. Rowe (Eds.), *Energy risk management* (pp. 223–245). London: Academic Press.

Shetterly, C. (1988, October 14). Scientists find two volcanoes at Yucca. *Las Vegas Review-Journal*, pp. 8B,

Söderberg, O. (1989, June). *Information strategies in Sweden concerning the disposal of nuclear waste.* Paper presented at the Advisory Group Meeting on Public Understanding of Radioactive Waste Management Issues, Vienna.

Special report: Nuclear waste. (1994, Second Quarter). *Nuclear Energy*, pp. 2, 18–35.

Svensk Kärnbränslehantering AB—Swedish Nuclear Fuel and Waste Management Co. (SKB) (1989). *Handling and final disposal of nuclear waste: Programme for research development and other issues.* Stockholm: Author.

_____ . (1992). *Activities, 1991.* Stockholm: Author.

Swainston, H.W. (1991). The characterization of Yucca Mountain: The status of the controversy. *Federal Facilities Environmental Journal,* **Summer,** 151–160.

Thurber, J.A. (1994, March 1). *Report on selected published works and written comments regarding the Office of Civilian Radioactive Waste Management Program, 1989–1993* (Prepared for Secretary Hazel R. O'Leary, U.S. Department of Energy). Washington, DC: Center for Congressional and Presidential Studies, School of Public Affairs, American University.

Toro, T. (1992). Germany forces pace on nuclear dump. *New Scientist, 136*(1842), 10.

U.S. Congress, Office of Technological Assessment (OTA). (1982). *Managing the nation's commercial high-level radioactive waste.* Washington, DC: GPO.

_____ . (1991, February). *Complex cleanup: The environmental legacy of nuclear weapons production* (OTA–O–484). Washington, DC: GPO.

U.S. Department of Energy (DOE). (1978, February). *Report of the task force for review of nuclear waste management.* Washington, DC: GPO.

_____ . (1984, December). Advisory Panel on Alternative Means of Financing and Managing Radioactive Waste Facilities. *Managing nuclear waste—A better idea* (A report to the U.S. Secretary of Energy). Portland, OR: Author.

_____ . (1986a). Office of Civilian Radioactive Waste Management, *Recommendation by the Secretary of Energy of candidate sites for site characterization for the first radioactive-waste repository* (DOE/S–0048). Washington, DC: GPO.

_____ . (1986b). Office of Civilian Radioactive Waste Management. *A multiattribute utility analysis of sites nominated for characterization for the first radioactive-waste repository: A decision-aiding methodology* (DOE/RW– 0074). Washington, DC: GPO.

_____ . (1987). Office of Civilian Radioactive Waste Management. *OCRWM mission plan amendment* (DOE/RW–1028). Washington, DC: GPO.

_____ . (1989, November 29). Office of Civilian Radioactive Waste Management, *Report to Congress on reassessment of the civilian radioactive waste management program* (DOE/RW–0247). Washington, DC: GPO.

_____ . (1990). Office of Civilian Radioactive Waste Management. *Integrated data base for 1990: U.S. spent fuel and radioactive waste inventories, projections, and characteristics.* Washington, DC: GPO.

_____ . (1993, October 14). Yucca Mountain Site Characterization Office. *Yucca Mountain project public participation program and guidance* (draft). Washington, DC: GPO.

U.S. General Accounting Office (GAO). (1985). *Status of the DOE's implementation of the Nuclear Waste Policy Act of 1982* (GAO/RCED–85–116). Washington, DC: Author.

_____ . (1987a). *Nuclear waste: Institutional relations under the Nuclear Waste Policy Act of 1982* (GAO/RCED–87–14). Washington, DC: Author.

_____ . (1987b). *Nuclear waste: Status of DOE's nuclear waste site characterization activities* (GAO/RCED–87–103–FS). Washington, DC: Author.

_____ . (1987c). *Nuclear waste: Status of DOE's implementation of the Nuclear Waste Policy Act* (GAO/RCED–87–17). Washington, DC: Author.

_____ . (1990, October). *Nuclear energy: Consequences of explosion of Hanford's single-shell tanks are understated* (GAO/RCED–91–34). Washington, DC: Author.

_____ . (1991, August). *Nuclear waste: Hanford single-shell tank leaks greater than estimated* (GAO/RCED–91–177). Washington, DC: Author.

_____ . (1992, May). *Nuclear waste: DOE's repository site investigations, a long and difficult task* (GAO/RCED–92–73). Washington, DC: Author.

_____ . (1993, May). *Nuclear waste: Yucca Mountain project behind schedule and facing major scientific uncertainties* (GAO/RCED–93–124). Washington, DC: Author.

_____ . (1994a, July 27). *Comments on the draft statement of work for the financial and management review of the Yucca Mountain project* (GAO/RCED–94–258R). Washington, DC: Author.

_____ . (1994b, August). *Nuclear waste: Foreign countries' approaches to high-level waste storage and disposal* (GAO/RCED–94–172). Washington, DC: Author.

_____ . (1994c, September). *Nuclear waste: Comprehensive review of the disposal program is needed* (GAO/RCED–94–299). Washington, DC: Author.

U.S. Secretary of Energy. (1993, November). Advisory Board Task Force on Radioactive Waste Management. *Earning public trust and confidence: Requisites for managing radioactive waste.* Washington, DC: Author.

van Konynenburg, R.A. (1991). Gaseous release of carbon-14: Why the high-level waste regulations should be changed. In *High-level radioactive waste management: Proceedings of the second annual international conference* (pp. 313–319). La Grange Park, IL: American Nuclear Society and American Society of Civil Engineers.

Walker, J.S. (1992). *Containing the atom: Nuclear regulation in a changing environment, 1963–1971.* Berkeley: University of California Press.

Walters, C.J. (1986). *Adaptive management of natural resources.* New York: MacMillan.

Weart, S.R. (1988). *Nuclear fear: A history of images.* Cambridge, MA: Harvard University Press.

Wiltshire, S. (1986). Public participation in Department of Energy high-level waste management programs. *Tennessee Law Review, 53,* 540–557.

Younker, J.L., Andrews, W.B., Fasano, G.A., Herrington, C.C., Mattson, S.R., Murray, R.C., Ballou, L.B., Revelli, M.A., Ducharme, A.R., Shephard, L.E., Dudley, W.W., Hoxie, D.T., Herbst, R.J., Patera, E.A., Judd, B.R., Docka, J.A., & Rickertsen, L.R. (1992). *Report of early site suitability evaluation of the potential repository site at Yucca Mountain, Nevada* (SAIC–91/8000). Washington, DC: U.S. Department of Energy.

Yucca Mountain Socioeconomic Study Team. (1993). *The State of Nevada, Yucca Mountain Socioeconomic Studies, 1986–1992: An annotated guide* (NWPO–SE–056–93). Carson City, NV: NWPO.

Ziemer, P.L. (1992, August 12). Letter to William Rosenberg describing DOE's objections to EPA's proposed standard for disposal of high-level nuclear waste (40 CFR 191).

About the Book and Authors

When Congress passed the Nuclear Waste Policy Act of 1982, it directed the Department of Energy to locate, study, license, and develop a deep underground repository for high-level nuclear wastes. As the authors of this study show, by 1987 the program was in shambles, beset by opposition from every state that had a potential storage site. Congress passed amendments to the original legislation that designated Yucca Mountain, Nevada, as the only site for study and development.

The authors trace the evolution of the political and social turmoil created by this difficult site-selection process, looking at the history of the nation's repository program, the nature of the public's concerns, and the effects of intergovernmental conflict. They also examine how other countries have addressed similar problems. Turning to a promising development—a dry-cask storage method judged by the Nuclear Regulatory Commission to be safe for a century or more—they urge a full reassessment of the nation's high-level nuclear waste policies and of existing DOE programs.

The book concludes with carefully considered recommendations for a new national policy for the storage of hazardous nuclear waste. Everyone concerned about nuclear waste and how it should be managed at the federal, state, and local levels will find valuable information in this in-depth study of the issues at hand.

James Flynn is a Senior Research Associate at Decision Research in Eugene, Oregon. He served as the Yucca Mountain socioeconomic studies project manager from 1986–1992, and since then as the project director. He has conducted socioeconomic research for federal agencies, states, private companies, and local governments on a number of industrial projects including the siting, licensing, and study of nuclear power and nuclear waste facilities. Recent publications on nuclear issues include articles in *Environment, Science, Risk Analysis, Issues in Science and Technology, Energy Studies Review,* and *Forum for Applied Research and Public Policy.* Dr. Flynn holds a Ph.D. from the University of Washington.

James Chalmers, a partner in the consulting practice of Coopers & Lybrand, L.L.P., is located in Phoenix, Arizona. He was the project director of the Yucca Mountain socioeconomic studies from 1986–1992 and currently is a senior consultant to the study team. He received a Ph.D. in economics from the University of Michigan.

Doug Easterling is the Officer for Research and Evaluation at The Colorado Trust, a Denver-based foundation that funds health-related initiatives in Colorado. His role at The Colorado Trust is to develop and monitor research projects that evaluate the effectiveness of initiatives in areas such as teenage pregnancy, health promotion, and community development. He received his doctorate in Public Policy and Management at the Wharton School of the University of Pennsylvania. Over the past 12 years, Easterling has conducted research in the fields of risk perception, psychometrics, health psychology, and environmental and health policy.

Roger Kasperson, Professor of Government and Geography and Senior Researcher at Clark University, earned his Ph.D. from the University of Chicago. Dr. Kasperson is co-author or co-editor of many books on technological hazards, risk communication, public responses to risk, radioactive wastes, and global environmental change. He has directed numerous research projects involving these topics, as well as on ethical and policy issues in risk management, the social amplification of risk, and social visions of a sustainable society. Dr. Kasperson has worked as consultant or advisor to public and private agencies on energy and environmental issues and is a Fellow of the American Association for the Advancement of Science. He is currently serving as Provost at Clark University.

Howard Kunreuther is the Cecelia Yen Koo Professor of Decision Sciences and Public Policy, as well as Co-Director of the Wharton Risk Management and Decision Processes Center at the University of Pennsylvania. His current research examines the role of insurance compensation, incentive mechanisms, and regulation as policy tools for dealing with technological and natural hazards. He is author and co-author of numerous scientific papers concerned with risk and policy analysis, decision processes, and protection against low-probability/high-consequence events, as well as many books and monographs. Dr. Kunreuther has a Ph.D. in economics from the Massachusetts Institute of Technology.

C.K. Mertz is a researcher and data analyst with Decision Research in Eugene, Oregon. She has been involved with the State of Nevada socioeconomic research on the high-level nuclear waste repository proposed for Yucca Mountain, Nevada, since 1986. Her experience includes social, economic, and policy research, data analysis, and database management. She is a co-author of several articles and published reports.

Alvin Mushkatel is Professor of Public Affairs and Director of the Office of Hazard Studies at Arizona State University. He has authored and co-authored several books and articles on natural and environmental hazards policies. He recently co-edited a book on nuclear waste disposal policy (Greenwood Press). His current research examines the importance of public trust and stakeholder involvement in policies designed to clean up or reduce risk resulting from the use or production of hazardous technologies. Dr. Mushkatel serves on two National Research Council committees: one that examines the nation's efforts to eliminate the Army's chemical weapons stockpile, and one that analyzes the costs to decontaminate and decommission the nation's uranium enrichment plants.

K. David Pijawka is a professor in the school of Architecture and Planning at Arizona State University. He received a Ph.D. in geography from Clark University and has written numerous publications on the subjects of nuclear technology and environmental hazards.

Paul Slovic, President of Decision Research and Professor of Psychology at the University of Oregon, received a Ph.D. in psychology from the University of Michigan. He studies human judgment, decision making, and risk analysis. Dr. Slovic publishes extensively and maintains research relationships with colleagues throughout North America, Europe, and Asia. He is past President of the Society for Risk Analysis and in 1991 received its Distinguished Contribution Award. He also serves on the Board of Directors for the National Council on Radiation Protection and Measurements. In 1993 he received the Distinguished Scientific Contribution Award from the American Psychological Association.

Lydia Dotto is a freelance writer who has specialized in science and environmental issues for 23 years. She has authored 10 previous books, including several on environmental issues commissioned by scientific groups and organizations. She is the former science reporter for Canada's national newspaper, *Globe and Mail,* and former executive editor of Canadian Science News Service. She has won numerous awards, including the Royal Canadian Institute's Sandford Fleming Medal for outstanding achievement in promoting understanding of science among the Canadian public.

Author Index

Ahagen, H., 84
Ahearne, J.F., 27, 29
Ahlström, P.-E., 88

Blowers, A., 85
Broad, W., 23, 109

Carter, L.C., 21, 37, 38, 40, 45
Carter, L.J., 31
Carver, S., 85
Clark, R..L., 42
Chalmers, J., 76
Colglazier, E.W., xii, 25, 34, 36, 37
Cook, B.J., 38, 64, 85, 86

Dantico, M., 45
Davis, J.A., 40
Desvousges, W.H., 76
DOE, 13, 20, 22, 23, 39, 46, 50, 103

Edison Electric Institute, 53
Emel, L.J., 38, 64, 85, 86
Erikson, K., xii, 25, 77

Fernie, J., 85
Fischhoff, B., 77
Flynn, J., 27, 50, 51
Fiske, S.T., 77

Galpin, F.L., 42
General Accounting Office (GAO), 22, 26,
    28, 44, 46, 53, 84, 106, 107
German, J., 52
Gesell, G., 76
Graham, B., 43
Greenwood, M., 74

Hansson, S.O., 56, 57
Hardin, E., 84
Healy, M., 80
Herzik, E.B., 49

Holdren, J.P., 20
Holmberg, S., 88

Jacob, G., 29, 36, 38, 46, 47, 48, 49

Kasperson, J.X., 59
Kasperson, R.E., 27, 34, 38, 59, 64, 78, 79,
    85, 86
Keeney, R.L., 39, 41, 61
Kraft, M., 38, 45, 46
Kraus, N., 76
Kunreuther, H., 27, 76

Ladanyi, B., 91
Langum, R.B., 34, 36
Layman, M., 76
Lenssen, N., 20
Lichtenstein, S., xii, 77
Lowry, D., 85

Merkhofer, M.W., 39, 41
Mertz, C.K., 50, 51
McClelland, G., 74
McCombie, C., 91
McKay, B., 36, 49
McKinley, I., 91
Mountain West, 75
Mushkatel, A., 45, 47, 49, 50

National Research Council/National Acad-
    emy of Sciences (NRC/NAS), 6, 24,
    55, 56
Nevada Nuclear Waste Project Office
    (NWPO), 103, 105
Nuclear Waste Strategy Coalition, 105
Nuclear Waste Technical Review Board
    (NWTRB), 28, 29, 53, 59, 88, 90, 91,
    92

Openshaw, S., 85
OTA, 46, 48, 53

Subject Index